Women in Engineering and Science

Series Editor
Jill S. Tietjen
Greenwood Village, Colorado, USA

More information about this series at http://www.springer.com/series/15424

Takoi Khemais Hamrita

Editor

Women in Precision Agriculture

Technological breakthroughs, Challenges
and Aspirations for a Prosperous
and Sustainable Future

 Springer

Editor
Takoi Khemais Hamrita
Driftmier Engineering Center
University of Georgia
Athens, GA, USA

ISSN 2509-6427 ISSN 2509-6435 (electronic)
Women in Engineering and Science
ISBN 978-3-030-49246-5 ISBN 978-3-030-49244-1 (eBook)
https://doi.org/10.1007/978-3-030-49244-1

This Springer imprint is published by the registered company Springer Nature Switzerland AG
The registered company address is: Gewerbestrasse 11, 6330 Cham, Switzerland

To the loving memory of my father Khemais Hamrita who was my role model, cheerleader, anchor, and guide, and who instilled in me a hunger for knowledge and a burning desire to make a difference
To my mother Jamila Dardour who taught me the meaning of hard work, selflessness, and patience
To you Ayman for helping me grow and persevere in ways I never thought possible

Preface

I write this book preface during the COVID-19 crisis, while social distancing in my home. To me, this crisis has revealed a number of invaluable lessons to the world. The two lessons most important to the context of this book are: (1) science and technology are vital to our collective future, and (2) problems of the future can't be solved in isolation, we must work together. As the founder of the Global Women in STEM Leadership Summit, I have the privilege to interact with some of the most brilliant women in STEM (Science, Technology, Engineering, Math). Through these interactions and through the summit programs, I have learned that we are far from achieving equity in STEM. We, women in STEM, are still severely under-represented and our efforts are to a great extent marginalized. This is not only detrimental to our well-being as women, but it is also detrimental to the advancement and preservation of our collective future as a humankind.

This book is a small contribution towards promoting the efforts of women in my technical field, precision agriculture. Precision agriculture is the application of technology to study, assess, and control the agricultural process and its inputs/outputs. To me, this field is a natural fit for women. Our land, our environment, and natural resources are stressed to the max. New solutions to steer us in a more positive and sustainable direction are becoming more vital every day. In my conversations with all the women I meet through the summit, one thing has become increasingly clear: women are driven by the common good, they like pursuing passions, and they care about their communities, the environment, animals, etc. Precision agriculture is inherently designed to optimize the use of natural resources and reduce waste, and to contribute to our well-being and that of the environment and the animals within it. A natural fit.

As you will see through the pages of this book, the passion and care of the contributing women authors are evident in every chapter. The chapters present cutting-edge research in almost every area of precision agriculture. I am particularly pleased that our book is unique in addressing both land and animal agriculture. I'm also delighted that the women authors represent various geographic areas around the world, and many disciplines and career paths and stages. I hope you will find this book not only informative and helpful in your own precision agriculture journey but also inspiring. We, women in STEM, are a vital intellectual, emotional, and spiritual resource for vital new global paradigms such as precision agriculture. Our collective future as a species depends on women in STEM being represented, being seen, and being heard!

Athens, GA, USA   Takoi Khemais Hamrita

Acknowledgments

This book wouldn't have come into existence were it not for Jill Tietjen's invitation to contribute to her Women in Engineering and Science book series. Thank you Jill for including me in your very special series and for being an amazing role model. I would also like to thank Springer for their invitation to create this book and for all the support they have given me. In particular, special thanks to Mary James, Hemalatha Velarasu, Zoe Kennedy, Brian Halm, and Mario Gabriele for their assistance with the book.

I am very grateful to the many women authors who have contributed to this book. They represent various geographic areas, and many disciplines and career paths and stages. I'm indebted to them for the huge time and effort they invested to produce high quality relevant content.

I would also like to thank my undergraduate research assistants Kaelyn Deal, Selyna Gant, Haley Selsor, Taylor Ogle, and Amanda Yi for their assistance with all aspects of putting this book together. In particular, I'm grateful for the thorough literature searches they conducted and the quality of references they produced.

I'm fortunate to have created the Global Women in STEM Leadership Summit. This community of very successful women and men has opened my eyes to the difficulties and inequities women in STEM face, and hence has fueled my interest in highlighting, advancing, elevating, and promoting women in STEM fields. I am very grateful to this powerful community.

Last but not least I would like to thank my personal tribe. I am fortunate to have a very loving family who believes in me and supports all my professional endeavors and pursuits. I am forever indebted to my late father Khemais Hamrita who dedicated so much of his wisdom, time, and energy to supporting, encouraging, and elevating me. I am able to empower others because he empowered me.

Contents

Chapter 1
Precision Agriculture: An Overview of the Field and Women's Contributions to It

Takoi Khemais Hamrita, Kaelyn Deal, Selyna Gant, and Haley Selsor

Contents

1.1   Introduction

By 2050, the world's population will exceed 9 billion people (Pandey 2018). According to some projections, to feed the world's population, the agricultural sector must increase production by 70% (Yun et al. 2017). This urgent need for

T. K. Hamrita (✉) · K. Deal · S. Gant · H. Selsor
School of Electrical and Computer Engineering, College of Engineering,
University of Georgia, Athens, GA, USA
e-mail: thamrita@uga.edu

© Springer Nature Switzerland AG 2021
T. K. Hamrita (ed.), *Women in Precision Agriculture*, Women in Engineering
and Science, https://doi.org/10.1007/978-3-030-49244-1_1

increased productivity, coupled with the need for reduced cost, more optimal use of resources, and reduced impact on our environment, has made it imperative to create and adopt new ways of farming. Many industries around the world are stepping up to meet the expected demand and supply. In agriculture, many researchers have turned to technology to aid food production.

Historically, agriculture has not benefited from systems thinking nor the tools and technologies that have been developed to control and optimize systems in other industries, such as manufacturing automation, process control, and aerospace. In these types of systems, in order to control and optimize performance, the outputs along with other important system variables are monitored, and their measurements are used to determine the inputs that would produce the desired performance. Precision agriculture is about viewing and treating the agricultural process as a system and incorporating information available from all its parts to improve its performance. In order to do so, new methods, tools, processes, and technologies had to be developed to enable the observation and measurement of important variables (Phadikar et al. 2012), facilitate the study and assessment of these variables to extract relevant information and knowledge (Castle et al. 2015), and use this knowledge to control the agricultural process and its inputs/outputs (Shobha et al. 2008). The goal is to create farming practices that respond precisely to the spatially and temporally varying needs of land and livestock, therefore optimizing yield while reducing cost and environmental impact. In other words, precision agriculture is about listening to the needs of the land, the animals, the environment, the farmers, and the consumers and doing what it takes to respond to these needs. It's about being holistic and tuning in to all parts of the system to make optimal management decisions.

Precision agriculture is a complex research and development field that lies at the interface of various disciplines and technological advances. Precision agriculture began to develop in the twentieth century with the help of researchers. As research started to release results, visionaries and scientists continued this trend and created precision agriculture (Srinivasan 2006). The trends arising in the research and development of precision agriculture include data mining, machine learning, Big Data, Small Data, data analytics, geographic information systems (GIS), Global Positioning System (GPS) and GPS auto guidance equipment, unmanned aerial vehicles (UAVs), Internet of Things (IoT), remote sensing, smart sensor networks, variable rate technology, nanotechnology, and robotics.

This book is a compilation of contributions, breakthroughs, and impactful research done by leading female researchers and scholars from various fields and from around the world toward making precision agriculture a reality. Tables 1.1a and 1.1b show the diverse technical, career paths and stages, and geographic backgrounds of the authors of this book. Tables 1.2a and 1.2b show examples of patents by leading women researchers in precision agriculture. Additionally, women authors or coauthors of research referenced in this chapter are highlighted. All these women are creating new technological advances that are revolutionizing agriculture and providing innovative solutions to some of today's most challenging global food problems, paving the way for a smarter, more precise, more efficient, and more profitable agriculture for the twenty-first century. The chapters in this book present a holistic view of the field, highlighting relevant technologies, decision-making strategies, practices, applications, economics, opportunities, and challenges for both land and livestock applications. This is the only known book focused on advances in precision agriculture for both land and livestock, led by women researchers and scholars, hence providing a unique woman's perspective in a field primarily dominated by men.

Table 1.1a Author and coauthor contributors to Chaps. 1, 2, 3, 4, and 5

Author name	Title and affiliation	Country
Chapter 1: Precision Agriculture: An Overview of the Field and Women's Contributions to It		
Takoi Hamrita	Professor and Inaugural Chair of the School of Electrical and Computer Engineering at the University of Georgia (UGA)	United States
Kaelyn Deal Selyna Gant Haley Selsor	Undergraduate Research Assistant University of Georgia	United States
Chapter 2: Sensing Technologies and Automation for Precision Agriculture		
Man Zhang	College of Information and Electrical Engineering, China Agricultural University	China
Ning Wang	Professor at Oklahoma State University, Department of Biosystems and Agricultural Engineering	United States
Liping Chen	Beijing Research Center of Intelligent Equipment for Agriculture, Beijing Academy of Agriculture and Forestry Sciences	China
Chapter 3: Perspectives to Increase the Precision of Soil Fertility Management on Farms		
Joann Whalen	Professor at McGill University and Adjunct Professor at Gansu Agriculture University	Canada
Chapter 4: Toward Improved Nitrogen Fertilization with Precision Farming Based on Sensor and Satellite Technologies		
Heide Spiegel	Senior Scientist at Agency for Health and Food Safety (AGES) and Department for Soil Health and Plant Nutrition	Austria
Taru Sandén	Postdoctoral Researcher, Department for Soil Health and Plant Nutrition at AGES	Austria
Laura Essl	Dipl.-Ing., Institute of Geomatics at the University of Natural Resources and Life Sciences	Austria
Francesco Vuolo	Senior Scientist, Institute of Geomatics, University of Natural Resources and Life Sciences	Austria
Chapter 5: Precision Weed Management		
Sharon Clay	Distinguished Professor, Plant Science Department at South Dakota State University	United States
Anita Dille	Professor in Weed Ecology and Assistant Head for Teaching in the Agronomy Department at Kansas State University	United States

Table 1.1b Author and coauthor contributors to Chaps. 6, 7, 8, 9, and 10

Author name	Title and affiliation	Country
Chapter 6: Precision Irrigation: An IoT-Enabled Wireless Sensor Network for Smart Irrigation System		
Sabrine Khriji	Research Assistant at the Chemnitz University of Technology and University of Sfax	Germany, Tunisia
Dhouha El Houssaini	Research Assistant at the Chemnitz University of Technology and University of Sfax	Germany, Tunisia
Ines Kammoun	Professor at the National Engineering School of Sfax	Tunisia
Olfa Kanoun	Professor at the Chemnitz University of Technology	Germany
Chapter 7: Women Farmer-Breeder Partnerships in Plant Breeding, Seed, and Food Innovations: Experiences from Tigray, Northern Ethiopia		
Fetien Abay Abera	Vice President for Research and Community Services at Mekelle University	Ethiopia
Chapter 8: Synthesis of a Research Program in Precision Poultry Environmental Control Using Biotelemetry		
Takoi Hamrita	Professor and Inaugural Chair of the School of Electrical and Computer Engineering at the University of Georgia	United States
Taylor Ogle Amanda Yi	Undergraduate Research Assistant University of Georgia	United States
Chapter 9: Advancement in Livestock Farming through Emerging New Technologies		
Jarissa Maselyne	Researcher at Flanders Research Institute for Agriculture, Fisheries, and Food (ILVO)	Belgium
Chapter 10: The Impact of Challenges and Advances of Bush Internet Connectivity for Women in Agriculture in Queensland, Australia		
Rachel Hay	Social Scientist and Lecturer in Marketing at James Cook University	Australia

1.2 Precision Agriculture in Crop Production: Enabling Technologies and Applications

Precision agriculture makes use of the understanding of variability within land and crops to implement spatially and temporally variable application of agrochemicals. As one analyst suggests, a way to view farming is as a branch of matrix algebra that juggles the variable inputs to a farm and the required analysis to understand the quantity those inputs are needed at (Technology Quarterly 2016). In addition to informing agricultural input decisions, precision agriculture can suggest to the farmers the right crop based on their site-specific parameters (Pudumalar et al. 2017). Being able to measure variability within a field and apply inputs accordingly requires a number of enabling technologies.

Table 1.2a A sample of US female patent holders in precision agriculture

Name	Title and affiliation	Patent #	Patent title
Takoi Hamrita et al.	Professor at the University of Georgia	6525276	Crop yield monitoring system
Sarah L. Schinckel et al.	Manager at Deere and Company	10150483	Settings manager – distributed management of equipment and display settings via centralized software application
Margaux M. Price et al.	Manager at Deere and Company	10150483	Settings manager – distributed management of equipment and display settings via centralized software application
Kimberly A. Salant et al.	CEO and Co-owner of Sentek Systems	9945828	Airborne multispectral imaging system with integrated navigation sensors and automatic image stitching
Lori J. Wiles et al.	Quantitative Agricultural Scientist at Colorado State University	9563852	Pest occurrence risk assessment and prediction in neighboring fields, crops, and soils using crowdsourced occurrence data
Shelley Haveman Wolff et al.	Principal Investigator, Microbiology at Luca Technologies	20140154727	Geobacter strains that use alternate organic compounds, methods of making, and methods of use thereof
Zarath Morgan Summers et al.	Researcher at University of Massachusetts Amherst	20140154727	Geobacter strains that use alternate organic compounds, methods of making, and methods of use thereof
Lynn David Jensen et al.	Title and affiliation unknown	US6701857B1	Depth control device for planting implement
Chloe Romier	Global Product Manager at GEOSYS SAS	9972058	Method for correcting the time delay in measuring agricultural yield
Cherish Bauer-Reich et al.	Assistant Professor of Engineering at NDSU Research Foundation	9964532	Biodegradable soil sensor, system, and method
Emily Rowan et al.	Title unknown, affiliated with Climate Corporation	US10028451B2	Identifying management zones in agricultural fields and generating planting plans for the zones

1.2.1   The Global Positioning System (GPS)

A cornerstone technology for precision agriculture is the Global Positioning System (GPS). The GPS, along with new sensor technology, has enabled the development of various types of maps that allow farmers to visualize their land, crops, and management in unprecedented ways (Yousefi and Razdari 2015). With this variability information, and visualization of their land, farmers can better manage their resources. The satellite-based GPS system was first created in the 1970s by the US

Table 1.2b A sample of European and Australian female patent holders in precision agriculture

Name	Title and affiliation	Patent #	Patent title
Montserrat Jurado Exposito et al.	Tenure Scientist at Consejo Superior de Investigaciones Cientificas CSIC	ES2245250A1, WO2005122755A1	Procedure for the discrimination of soil uses and the quantification of vegetable cover through teledetection with air photography
Francisca Lopez Granados et al.	Research Scientist, Laboratory Head at Consejo Superior de Investigaciones Cientificas CSIC	ES2245250A1, WO2005122755A1	Procedure for the discrimination of soil uses and the quantification of vegetable cover through teledetection with air photography
Nadia Shakoor et al.	Title and affiliation unknown	WO2018049189	Integrated field phenotyping and management platform for crop development and precision agriculture
Pramila Mullan et al.	Principal Director at Accenture	US9792557B2, AU2017228695B2	Precision agriculture system

Department of Defense. In the 1990s, agricultural engineers combined this technology with various yield sensing and data processing technologies to create crop yield maps. In the late 1990s, the US farmers began to use yield mapping technology to "see" bigger variations within their fields than they had ever imagined. Today, GPS receivers are common on farm equipment. Producers use GPS information to control and guide farm equipment and to map and monitor their farms. Without GPS as a reliable and affordable tool, it would have been hard for precision agriculture to become a viable and popular solution (National Museum of American History 2018).

1.2.2   Geographic Information Systems (GIS) (https://www. esri.com/en-us/what-is-gis/overview)

A geographic information system (GIS) is a framework for gathering, managing, and analyzing information. GIS integrates and organizes many types and layers of information using maps and 3D scenes. With this unique capability, GIS reveals deeper insights, patterns, connections, and relationships in the data, helping users make smarter decisions. Organizations around the world use GIS to make maps that communicate, analyze, and share information to solve complex problems.

Early applications of GIS in agriculture date back to the 1970s (Mulla and Khosla 2016). GIS technology is an integral part of PA as it is instrumental in creating maps that reflect variability within soil, crops, and yield across a field. These maps serve as the basis for making and executing optimal management decisions.

1.2.3   New Sensing Technologies Are the Backbone of Precision Agriculture

As it is the case for any type of system, being able to measure outputs of the system as well as other important variables is a prerequisite for controlling the system and obtaining the desired outcomes. For the application of precision agriculture in crop production, it is important to be able to measure properties of the soil and the crops, as well as the output or the yield as it is commonly referred to, in order to gain understanding of the spatial and temporal variability within fields. Some experts like Shannon Ferrell and her associates suggest that we are witnessing an information revolution in the agricultural sector as sensor technology and data analytics from other industries are now being applied to agricultural applications (Coble et al. 2018). According to Takoi Hamrita and colleagues, the need for these sensors stems from the necessity of real-time control in order to have high-quality agricultural production (Hamrita et al. 1996, 2000). Sensors also address labor shortages and meet regulatory constraints on safety and environmental responsibility (Hamrita et al. 1996, 2000). Availability of sensors and sensor data has driven a number of agricultural innovations such as variable rate technology, crop-specific yield monitors, UAVs, and GPS Guidance Systems (Castle et al. 2015). "Information became a new crop of the 21st century, making farmers more efficient and sustainable but increasingly technologically dependent" (National Museum of American History 2018). Precision agriculture would not be what it is today were it not for two key technological advances: remote sensing and wireless smart sensor networks.

In Chap. 2 of this book, Ning Wang (Oklahoma State University, USA), Man Zhang (China Agricultural University, China), and Liping Chen (Beijing Academy of Agriculture and Forestry Sciences, China) provide a detailed discussion of different sensor technologies geared toward sensing soil, root, and crop properties. The authors also discuss the various types of platforms that are used to meet sensing requirements of different PA applications including in-field sensor networks, ground mobile platforms, manned and unmanned aerial vehicles, and satellites. The chapter also gives two detailed examples of the use of sensing technologies in PA, namely, the use of wireless sensor networks for real-time soil property monitoring, and the use of a ground-based phenotyping platform to evaluate peanut canopy architecture.

Remote Sensing

Farmers gather remotely sensed information by using planes, UAVs, and low earth orbital satellites passing over their land (Technology Quarterly 2016). Airborne instruments attached to planes can measure plant coverage and make distinctions between weeds and crops. This distinction allows autonomous machinery to remove weeds without damaging or mistakenly removing valuable crops. Small satellites use multispectral imagery that observes plant absorption of varying wavelengths emitted by the sun (Technology Quarterly 2016). Recent technological developments in aerospace engineering have led to Low-Altitude Remote Sensing systems. These systems allow aerial images to be taken at low altitudes using unmanned aerial systems (UAS) (Zhang and Kovacs 2012), also referred to as unmanned aerial vehicles (UAV). An unmanned low-altitude imaging system is more accessible and affordable to producers than the more expensive manned aerial imaging systems or satellite imagery. In Zhang and Kovacs (2012), the authors provide a review of recent studies in the application of UAS imagery for PA. The authors analyze and discuss results of these studies and limitations of UAS in agriculture. Topics discussed include remote sensing, small UAS and environmental studies, limitations of UAS, platforms and cameras, UAS image processing, issues with aviation regulations, future application of UAS in PA, advancements in UAS, methods of data extraction from UAS imagery, and attracting farmer interest. Ana Isabel de Castro, Francisca López-Granados, Maggi Kelly, and colleagues indicate that UAVs can supplement small satellites with ultrahigh spatial resolution data to distinguish weeds from crops and finely tune site-specific weed management (Gómez-Candón et al. 2013; Peña et al. 2013), as well as detect spatial variability in yield using smart yield detection technology (Vellidis et al. 2001). UAVs have the potential to be used on all farms in the world (Zhang and Kovacs 2012) and are proving to be versatile in unexpected ways. For example, In Japan, UAVs are used for more than imaging as they are used to scare birds, spray areas, protect against theft, aid in the creation of fields' maps, and monitoring the evenness of germination and analysis of all the necessary nutrients to plant's availability over large areas (Yun et al. 2017).

Wireless Smart Sensor Networks

Recent innovations have made it possible for sensors to be smaller and cheaper and to be made with computer components according to Loredana Lunadei and colleagues (Ruiz-Garcia et al. 2009). Advances in wireless communication and digital circuits have made it possible to build wireless smart sensor networks. These sensors are called smart sensors as they are capable of wireless communication with each other and with other parts of the system, data processing, and computing (Hamrita et al. 2005). In the past decade, wireless sensors have been subject to continuous innovation, allowing them to complete increasingly complex tasks (Ivanov et al. 2015). Smart sensors are multidisciplinary, monitoring yield, inputs, and interactions between machines and crops. They integrate with other innovative technology such as autonomous machinery and have been widely accepted by farmers. The role of wireless sensor networks in agriculture has become vital as part of the precision farming initiative according to Kriti Bhargava and colleagues (Ivanov et al. 2015). Wireless communication platforms are developing to mitigate the need for wiring harnesses and maintenance and increase mobility. They allow sensor applications in remote or dangerous locations to be monitored. Additionally, wireless networks can take advantage of and utilize cellular phones, radios, and Global Positioning Systems. Sensor networks can be combined with data mining techniques to map behavioral patterns for different crops. Tatiana Gualotuna and colleagues suggest that their data can be used to create the most effective and productive management plan for specific crops (Rodríguez et al. 2017). In Ruiz-Garcia et al. (2009), Loredana Lunadei and her colleagues provide a review of wireless sensor technologies and communication systems such as wireless sensor networks (WSN) and radio frequency identification, and the application of these technologies in agriculture is discussed. These applications include fire detection, climate monitoring, crop canopy influence, climate influence, farm machinery, pest control, viticulture, irrigation, greenhouses, livestock, and cold chain monitoring and traceability. Future trends are also discussed.

Sensor Applications

Soil Sampling

Soil sampling to study its properties is not new. In the 1970s, groups of soil scientists studied the spatial variability of soil moisture and hydraulic properties and its use to improve the precision of soil mapping (Mulla and Khosla 2016). In the 1980s, building on these studies, research was done to map phosphorus levels in soil, and software was developed to automatically classify and map soil fertility sampling data into fertilizer management zones. This was the first combined use of geostatistics and GIS for precision farming (Mulla and Khosla 2016). This was also most likely the first real application of PA (Srinivasan 2006).

Several sensing technologies have been developed since to sense various properties of the soil. In Adamchuk et al. (1999), the authors discussed the development of an automated soil sampling system that extracts soil at a designated depth and measures the pH every 8 s. This allows farmers to insightfully administer fertilizers to adjust the pH of specific parcels of land instead of the whole field, conserving resources and minimizing costs. In Kühn et al. (2008) Sylvia Koszinski and colleagues explore the viability of using the spatial variability of electrical conductivity in soil as a digital soil mapping tool. This method proved to offer a more detailed and lower level look at soil properties rather than traditional geological maps.

Applying PA techniques to soil fertility management will optimize agricultural production while protecting the environment. In Chap. 3 of this book, Joann Whalen (McGill University, Canada) discusses soil fertility and evaluation methods of soil nutrient status to aid in the selection of the right source and amount of fertilizer, and the best time and place to add fertilizers so the nutrients will be used efficiently by the crop. Joann makes the case for how technological advances, such as low-cost sensors, make it possible to apply nutrient-rich fertilizers to crops in smaller doses and with greater precision at the field scale, therefore avoiding nutrient losses from agroecosystems to surrounding environments. In Chap. 2 of this book, the authors discuss the use of wireless sensor networks for real-time soil property monitoring.

Weed Control

Precision weed management consists of optimizing inputs to reduce weed presence and improve crop yields. In Gómez-Candón et al. (2013), UAVs coupled with ground control points produced ultrahigh spatial resolution to locate weeds in their early phenological stages throughout a wheat field. The primary objective that this study supported is the creation of broad-leaved and grassweed maps in wheat crops for early site-specific weed management. In Peña et al. (2013), using UAVs, a six-band spectral camera, and object-based image analysis (OBIA), weed maps were made at the early phenological stage of both the crops and weeds themselves. The study aimed at creating highly detailed weed maps for early site-specific weed management to support Europe's recently passed legislation for more sustainable practices with pesticides.

In Chap. 5 of this book, Sharon Clay (South Dakota State University) and J. Anita Dille (Kansas State University) present a very thought-provoking discussion of site-specific weed management. In particular, they highlight opportunities for precision weed management, methods to collect and process information needed to implement precision management, current knowledge available on the topic, and challenges and recommendations for success. They indicate that no matter the method, understanding which weeds are present, location, density, and the appropriate control and timing are critical for success. They also argue that in the future, agronomists will need to have knowledge and skills beyond traditional weed science to include an understanding of state-of-the-art topics such as robotics as well as big data management and processing.

Nitrogen Fertilization

Nitrogen is a crucial nutrient for crops. Precision farming aims at adjusting nitrogen fertilization according to the needs of the plants while minimizing waste in the environment. In Haboudane et al. (2002), Louise Dextraze and colleagues predict the chlorophyll content of plants using remote sensing and hyperspectral imaging. This predictive tool aids in quantifying the amount of nitrogen dosage needed by plants to prevent excess nitrogen from running off into irrigation networks and waterways. In a study led by Jana Havránková, an analysis of passive and active ground-based remote sensing systems for canopy nitrogen management was conducted in winter wheat. The study revealed that both passive and active sensors uncovered nitrogen content of plants particularly well at the early growth stages, and succeeded in reducing the UK's nitrogen application by 15 kg/ha and in Slovakia by 1.5 kg/ha. Additionally, the UK reduced its residual nitrogen content in the soil by 52% (Havránková et al. 2007).

In Chap. 4 of this book, Heide Spiegel and Taru Sandén (Austrian Agency for Health and Food Safety, Austria), and Laura Essl and Francesco Vuolo (Institute of Geomatics, University of Natural Resources and Life Sciences, Austria) describe a study aimed at evaluating sensor and satellite technologies using field data collected under optimal conditions at different stages of nitrogen fertilization. Their work involved close cooperation between remote sensing experts, soil scientists, agronomists and practitioners, and their findings yielded productivity maps that serve as a basis for nitrogen fertilization maps for use by commonly applied farm machinery.

Irrigation Control

One of the greatest agricultural inputs is water and irrigation. Agricultural production requires a large amount of fresh water. As a result, agriculture must compete with other industries for potable water. Add in climate change, rainfall variation, and decreasing water tables, and there arises an obvious need for monitoring and optimizing water usage. Water management offers an effective solution to satisfy the world's growing demand for water (Dalezios et al. 2017). Irrigation systems can become more efficient with sensor networks, data science, and variable rate technology. Precision irrigation has great promise to improve the efficiency of water use. The importance of water availability estimation, agriculture water needs, and the necessity for monitoring drought conditions are essential to successful PA (Dalezios et al. 2017). Several studies in the literature have dealt with precision irrigation.

Gago et al. (2015) discusses how PA can be used for water stress management using unmanned aerial vehicles (UAVs) to monitor crop fields with high spatial and temporal resolution. It is suggested that these vehicles use thermal imagery to measure the difference between canopy and air temperatures or to measure chlorophyll fluorescence. The information gathered provides insight into water stress variability in a field and can also determine which crops are optimal for breeding purposes by measuring yield and stress tolerance. In a study led by Monica Diez, Cristina Moclan and colleagues (Díez et al. 2014), satellite imaging was used to monitor high-density cornfields and soya crop plots in northern Texas with the goal of making irrigation practices more efficient and effective. The study focused on high-frequency passes (every 2 days) over the test site in order to generate the best irrigation recommendations. The approach was considered successful by reducing irrigation costs and improving crop productivity (Díez et al. 2014). In Dalezios et al. (2019), the authors used Earth Observation data to measure water availability in fields in a drought-susceptible area in Greece. By monitoring water uptake of crops, rainfall, and drought potential, farmers can better manage their field and productivity.

In Chap. 6 of this book, Sabrine Khriji (Technische Universität, Germany, and University of Sfax, Tunisia), Dhouha El Houssaini (Technische Universität, Germany, and Technopark of Sousse, Tunisia), Ines Kammoun (National School of Engineers of Sfax, Tunisia), and Olfa Kanoun (Technische Universität, Germany) present a real-time IoT-based smart irrigation system. A number of wireless sensor nodes are used to monitor both soil moisture and temperature. Sensed data is transmitted to the gateway through the Queuing Telemetry Transport communication protocol. A Web interface and mobile application are provided to users to control the level of water in the soil in real time. Users can take immediate action to open or close the water pump through the mobile application.

Sensors in Harvesting

Sensors have also been developed to study the effects of harvesting and transportation processes on crops. For example, niche sensors needed for characterizing harvesting impact on soft-skinned berries and other sensitive crops are being developed by Takoi Hamrita and colleagues to improve upon the harvesting process and yield better products (Yu et al. 2011a, b). These sensors are vital for future understanding of the relationship between sensitive crops and machinery (Yu et al. 2011a, b).

Yield Mapping

Yield mapping is one of the most widely adopted PA technologies. Several studies in the literature discuss yield mapping and highlight the fact that it is a gateway technology to future adoptions of other complementary precision agricultural technologies. Yield maps reveal critical information about the status of the soil and the crops, therefore uncovering problem areas and providing insight into what areas need more resources or attention to increase their yield or productivity. For instance, in Aguilar-Rivera et al. (2018), the authors discussed yield monitoring of sugar crop fields in Mexico. Productivity or yield was measured with GPS and remote sensing technologies to show which land was highly productive and which needed improvement. Through yield mapping, land suitability was divided into three different levels, and to increase yield in less suitable areas, "agroecological management practices" were recommended for those areas. In another study of a peanut yield monitor developed at the University of Georgia (Vellidis et al. 2001), a yield map generated by the yield monitor uncovered problem locations such as a parasitic nematode infestations of peanuts and major yield variability. With the problem location known, more efficient and effective solutions could be applied saving both time and money. An important issue for yield monitors is to factor in and compensate for combine dynamics when interpreting yield data. It is recommended that to ensure accuracy with yield results, a yield reconstructive algorithm that takes into account combine dynamics should be developed. With proper installation, calibration, and operation of yield monitors, sufficient accuracy can be achieved in yield measurements to make site-specific decisions (Arslan and Colvin 2002).

1.2.4 Data Mining and Precision Agriculture

The use of numerous smart sensors, networks, and machines geared toward data collection has led the agricultural process to increasingly resemble a factory that produces mass data stemming from strict monitoring and control of a farm (Technology Quarterly 2016). Data harvested from PA can be used to maximize profits for a farmer. In a study led by Yan Ma, she and her colleagues found that

remote sensing data in particular has grown to be termed Big Data because of the sheer volume and the rate at which data is received for analysis (Ma et al. 2014). Data management paired with data analytics can allow site-specific fertilization for crops or tracking a specific beef cut from producer to consumer (Coble et al. 2018).

In order for precision technology to advance and become widely adopted, expansive databases with information regarding spatial variation within fields are needed. Data mining has become the common term for analysis to glean insightful conclusions and new information from available data. There are many methodologies to "mine" this data including machine learning, spike and slab regression analysis, time series analysis (Coble et al. 2018), relevance vector machines (Elarab et al. 2015), and variograms (Cameron and Hunter 2002; Kerry et al. 2010). These methods can be used in weather forecasting, crop yield prediction and selection, irrigation systems, crop disease prediction, and agricultural policy and trade (Coble et al. 2018). For example, to sift through multispectral and thermal infrared spatial imagery captured by a UAV, a relevance vector machine was used in Elarab et al. (2015) coupled with cross validation and backward elimination. The information gleaned from this data mining effort estimated chlorophyll concentrations and is used in other instances to predict future chlorophyll concentrations (Elarab et al. 2015; Haboudane et al. 2002). Another example of data mining involving the use of variograms to provide indications of the scale of variation within soil and crops is described in Cameron and Hunter (2002) and Kerry et al. (2010). From mining data in these manners, farmers get a better understanding of how many soil samples are needed to fully characterize a plot of land (Kerry et al. 2010). In Rodríguez et al. (2017), a wireless sensor network was developed and used to collect data for a rose greenhouse. This paper discussed data mining techniques to discover behavioral patterns of environmental conditions in the greenhouse.

Managing and processing big data still presents many challenges. In Skouby (2017), Danielle Skouby discusses the challenges and opportunities of Big Data being applied in agriculture and applied economics. Topics that are discussed include policy and legal implications, asymmetric information implications, sustainability, and traceability. In Ma et al. (2014), Yan Ma indicates that the technology receiving Big Data is not designed for quick and large amounts of data and data processing which reveals a gap of opportunity for technology to advance to meet data analytics level of development.

1.2.5   Robots and Variable Rate Technology

Variable rate technology allows inputs, such as water, fertilizer, and pesticides, to be analyzed by remote sensors and then altered by variable rate applicators in order to reduce waste and increase efficiency (Important tools to succeed in precision farming 2018). The success of variable rate technology depends on multiple factors including GPS, consumer perception, accuracy of the technology, yield mapping, and the perceived benefits by the farmer (O'Shaughnessy et al. 2019). In a study conducted in northern Texas, the viability of variable rate water irrigation on a center pivot was investigated using yield maps of the region (Marek et al. 2001). The study uncovered the need for detailed soil characterization data to determine the different management zones where irrigation will differ for select crops and rotations.

The introduction of robots on farms is a very recent event. Autonomous robots can improve farm safety by allowing farmers to remotely manage heavy machinery (Ozguven 2018). Robots also give farmers the ability to take a step back from their farm to address more pressing issues, and they eliminate human error (Important tools to succeed in precision farming 2018). Currently, field crops have reached highly autonomous levels, whereas specialty crops have not (Vasconez et al. 2019). Robots are being used in applications such as crop maintenance, harvesting, and livestock management. In Martelloni et al. (2016), the authors described the use of a robot for intra-row-flaming to control weed in maize. In Kutz et al. (1987), the authors described the use of a robot for planting seeds. The study indicated that the robot was able to determine the best work cell configuration, and the specialized gripper attachment on the robot successfully transported each seedling from the plug flat to the growing flat. An example of the use of robots in harvesting consisted of the development of a system that was able to distinguish between harvesting fruit and harvesting bunches of fruit (Tarrio et al. 2006). To robotically harvest small and delicate fruits that grow in bunches, the vision system was required to distinguish and locate each fruit within a bunch for adequate harvesting. This was done using a combination of stereoscopic cameras, laser diodes and structured lighting, and a 3D reconstruction of the fruit bunch's position. In Sáiz-Rubio et al. (2015) and Rovira-Más et al. (2015), a robot called VineRobot was designed to autonomously navigate a vineyard using a stereoscopic vision camera. The robot built real-time maps of vegetative growth in vines (using nitrogen content) and red grape maturity (using anthocyanin content). In livestock management, a robot termed the FeederAnt was designed and developed to be remotely controlled to feed piglets twice a day (Jørgensen et al. 2007). The robot placed food in different locations within a field in order to mitigate vermin and nutritional point leaching.

1.2.6 Nanotechnology and Precision Agriculture

Nanotechnology in agriculture is a recently developed area of research. Yet it has already proved a promising solution to agricultural problems such as improving agrochemical efficiency, food safety and sustainability, and environmental health (Pandey 2018). Nanotechnology is a multidisciplinary field that can be applied to PA, especially in the area of crop maintenance. For example, Surekha Duhan and colleagues show that viral disease detection kits for plants are beginning to stem from nanotechnology, and they increase a farmer's capability in fighting plant ailments in order to quickly detect and eradicate disease, while simultaneously reducing fertilizer input (Duhan et al. 2017). Nano-devices (Pandey 2018) can be used to sense and monitor farm conditions; for example, carbon nanotube devices can be used as sensors in extreme and critical bio-conditions. These devices can also be used to efficiently release nutrients in accurate doses for plant and livestock. In Duhan et al. (2017), the authors discuss nanotechnology and its impact on agriculture, and Pandey (2018) presents the possibilities and challenges of applying nanotechnology to agriculture from an Indian perspective. However, using nanotechnology in agriculture has raised some concerns among experts relating to the availability, synthesis, level of toxicity, health hazards, transportation challenges, and incongruity of regulatory structure (Pandey 2018). Rapid developments in nanopesticide research have motivated a number of international organizations to consider potential issues relating to the use of nanotechnology for crop protection (Kah and Hofmann 2014).

1.2.7 Breeding and Precision Agriculture

Breeding is important to agriculture because it takes the crop with the most productive characteristics and propagates it so that farmers achieve maximum efficiency. However, when there is large variability within a field, the ideal crop or phenotype might also vary. This is where precision agriculture is used to adapt to this variability. In Chap. 7, Feiten Abay reports on breeding efforts in the Tigray region of Ethiopia. In particular, she discusses the benefits of using a Participatory Plant Breeding approach and Participatory Varietal Selection. She then discusses the application of these approaches in barley production in Ethiopia.

1.3 Precision Agriculture in Animal Production: Enabling Technologies and Applications

As the global population has increased, the demand for food security has increased, and PA has acquired growing importance to optimize productivity to meet these global food demands. Livestock plays an important role in guaranteeing food security around the world and serves as an important contributor to the economy of many countries (Fort et al. 2017). It is already evident how PA practices have benefited crop production. Just as understanding variability in land is useful for crop PA, it is also useful for precision livestock farming (PLF). Applying PA to livestock involves taking feedback about well-being, and the effectiveness of management practices, directly from the animals, and doing so not only from the herd but also individually. According to Jessica Morris and colleagues, understanding the variability within livestock begins with a system in which livestock are managed as individuals or small groups rather than as a whole flock (Morris et al. 2012). Ilaria Fontana, Emanuela Tullo, and colleagues found that analyzing the well-being of livestock as individuals, and the combination of data obtained along with sophisticated algorithms that extract valuable information from it, provide valuable and rapid information to farmers as well as the management systems they use (Fontana et al. 2015). PLF offers support to farmers in their day-to-day management routine through the use of sensors, cameras, and microphones, and these have the potential to improve production (Fontana et al. 2015). PLF provides opportunities for increased productivity, increased profitability, and greater well-being of livestock.

PLF provides the opportunity to increase profits. Due to the use of sensors and technology to monitor the well-being of livestock, Jarissa Maselyne and colleagues suggest there can be detection of problems at an early stage, therefore preventing economic losses for the farmer (Maselyne et al. 2013). For instance, there is evidence that salivary cortisol is a promising biomarker for automatic monitoring of chronic heat stress which can be used by farmers to confirm that livestock are well (Baptista et al. 2013). In addition to detecting problems earlier, PLF increases profits by minimizing labor costs and resources used. Traditional livestock farming utilizes surveillance carried out by human operators which is time-consuming, costly, and prone to human variance (Thorton 2010). By using sensors to automatically collect data, and software to analyze, classify, and identify behaviors, there is precision that work conducted by humans lacks and that leads to profitability (Thorton 2010). Additionally, Harriet Wishart, Claire Morgan-Davies, and colleagues found that labor associated with feeding presents large costs to the farm business, but feeding with PLF reduces the required labor (Wishart et al. 2015). A mechanized feeding system provides supplementary feeding with the most efficient method, maximizing benefits and controlling costs (Wishart et al. 2015). PLF also contributes to greater well-being and health of animals. As the public has been informed of current treatment of livestock animals, concerns have been raised on how animal welfare can be guaranteed (Pastell et al. 2013). PLF meets these demands for better treatment of animals in different types of livestock farming including cattle, swine, poultry, and sheep.

Electronic control systems used with information and communication technology are the basic building blocks that enable precision livestock farming (Banhazi et al. 2012). Technological advances such as imaging and image analysis, GPS and GIS, feed sensors, air quality monitoring systems, augmented reality, and radio frequency identification systems have made it possible to manage livestock on an individual level. There have been studies in European countries and Australia where image analysis techniques were used to develop live animal measurement systems. For instance, in Banhazi et al. (2012), researchers developed a weighing system based on image analysis to determine the weight of individual or groups of pigs with acceptable precision. They did this by correlating dimensional measurements of the animals to their weights. Precision animal location monitoring allows researchers to evaluate pasture utilization, animal performance, and behavior (Turner et al. 2000). Turner indicates that GPS technology advances have allowed the development of lightweight collar receivers suitable for monitoring animal position at 5-min intervals. Position data can then be imported into a GIS to assess animal behavior characteristics and pasture utilization. Feed sensors can be used to precisely measure and control amounts of feed delivered to individual feeders (Benhazi et al. 2012). Air quality monitoring systems can be used to regulate the livestock environment since air quality has a significant effect on production efficiency (Benhazi et al. 2012). In a study led by Maria Caria in Italy, researchers used commercially available smart glasses (GlassUp F4 Smart Glasses) to scan QR codes that contained

Farm Information Sheets. These sheets contained "information about animals and feed stocks." The implementation of this technology was tested on a dairy sheep farm in Sardinia, Italy. Results indicated that smart glasses are useful for providing commonly needed livestock information and that the use of manual instead of vocal commands seemed to be the optimal solution, due to the presence of background noise in farming contexts (Caria et al. 2019). Low-frequency and high-frequency radio frequency identification (RFID) systems have become common in PLF applications. A recent development in RFID technology is ultrahigh-frequency RFID (UHF-RFID). A study conducted by Anita Kapun, Eva-Maria Holland, and Eva Gallman in Germany investigated the use of ultrahigh-frequency RFID in PLF (Adrion et al. 2018). Results indicated that UHF-RFID can be suitable for monitoring trough visits of pigs.

A notable comprehensive resource for PLF is a selection of 37 research papers from the 2014 European Association of Animal Production (EAAP) meeting (Halachmi 2015). This book is meant to facilitate cross-disciplinary discussion of animal sensing technology and related farm management. Participants came from the areas of animal sensing technology, agriculture, and animal sciences, and the topics addressed include an introduction to PLF; technologies for early detection of animal lameness; case studies about PLF's added value in pigs and cattle farms from the Netherlands, Spain, and Australia; genetics and the health of beef, calves, and heifers; automatic health detection in poultry and pigs; automatic health detection in cows; PLF in milk quality and contents; and rumen sensing as it relates to feed intake.

1.3.1 Cattle and Sheep PLF Applications

Assessing and alleviating disease and heat stress is a big concern for the cattle industry. In a study conducted by Marie-Madeleine Richard, Karen Helle Sloth, and Isabelle Veissier, they found that PLF provides the possibility of using real-time positioning to predict physiological and psychological states of the cows, which can indicate chronic stress or disease (Richard et al. 2015). If these issues are diagnosed in real time, they can be responded to quickly to ensure the health of the cow. In Halachmi (2015), GPS collars were placed on cattle to track the animal's positions at 5 min intervals. Analysis of this data provided insight into animal behavior characteristics and pasture utilization. In Richard et al. (2015), CowView technology from GEA Farm Technologies was used to detect the position of 350 cows to infer their activity. Through quantifying and analyzing the cow's activity, the circadian rhythm of cows was reported. Disruptions to this circadian rhythm indicate disease or chronic stress, providing an early detection system for farmers. There is research (Alsaaod et al. 2015) that suggests that using thermal radiation as a noninvasive technique to assess heat stress in cattle production is a good way to ensure that cattle are healthy without inflicting more stress by using invasive methods.

In Poulopoulou et al. (2015), the authors presented the development of a cloud-based software to use in the management of livestock farms. The software involves an app that monitors the farm through information of each cattle. This information can then be shared between farms in order to optimize farm management. In another study (Turner et al. 2000), Turner explains how PLF could lead to the understanding of how spatial and temporal variability of animals, forage, soil, and landscape features affect grazing behavior. This understanding provides potential to modify pasture management, improve efficiency of utilization, and maximize profits. PLF can also be used to predict calving events. Calving affects the physiological state of dairy cows significantly, and it is beneficial for farmers to know when these calving events are going to occur to ensure a sound birth and welfare of the dairy cow and calf (Zehner et al. 2019).

Several studies have demonstrated the effectiveness of using PLF within the sheep industry. For instance, within the British sheep farming industry, there is low record keeping on sheep farms due to the difficulty of gathering records and the lack of interest in using these records (Lima et al. 2018). Electronic identification (EID) provides farmers with a solution to these problems, impacting productivity by simplifying recording and retrieval of flock information and allowing data-driven management decisions (Lima et al. 2018). Another study (Morris et al. 2012) investigated the use of radio frequency identification technology in sheep farming in Australia. By using this technology, producers are able to monitor their sheep more efficiently and respond to issues faster. This has the potential to improve cost-effectiveness, labor efficiency, and consistency and reliability of information in the industry (Morris et al. 2012). Wishart and Morgan-Davies compare PLF to conventional methods of providing supplemental feedings during pregnancy of ewes in hill sheep systems. EID and automated weighing were used to sort the ewes into groups based on weight. More ewes treated with PLF methods reached standard feeding levels by the end of their pregnancy than the ewes treated with conventional methods, which continued to need supplemental feedings (Wishart et al. 2015). In a study led by Jessica Morris in Australia, the application of radio frequency identification technology in precision sheep management was investigated. This technology allowed producers to monitor sheep and improve the "efficiency of management and sheep welfare" (Morris et al. 2012).

In Chap. 10 of this book, Rachel Hay (James Cook University, Australia) discusses the inconsistencies in the adoption of PLF in the Australian beef industry, highlighting some of the existing barriers. In particular, she argues that marketing of PLF technologies has previously been aimed solely at men and that as technology diffuses into rural settings, it is modifying gender divisions and catalyzing productive partnerships in farming families. The chapter encourages stakeholders in the beef industry to consider women as leaders and decision-makers and highlights the valuable skills and attributes they bring to technology use and management.

1.3.2 Swine

For swine, the benefits of PLF are similar to those for cattle such as early detection of illness and species-specific events. According to Amy Miller and colleagues, there are certain behaviors that change in swine when disease is starting, and these changes are not easy to quantify and require lengthy observations by staff, which is impractical on a commercial scale (Matthews et al. 2016). Automated early warning systems use sensors to objectively measure swine behavior in order to identify any significant changes that might indicate stress or illness (Matthews et al. 2016). By identifying these issues early, they can be addressed to ensure swine are healthy. A common concern in swine farming is tail biting. It causes injury to swine, can lead to the spread of infection, and indicates a stressful living environment (Schrøder-Petersen and Simonsen 2001). In D'Eath et al. (2018), the authors highlight the potential for a 3D machine vision system that can provide early warning of tail biting on farms. This not only reduces the harm swine inflict on each other, it also indicates a stressful living condition before the tail biting which can be addressed to ensure the welfare of the swine. Feeding patterns among pigs also indicate possible disease or other health issues. In Jarissa Maselyne's study, she explains that the use of radio frequency identification technology can provide farmers with an early warning system for lameness and general illness with the potential to identify other issues (Maselyne et al. 2013). In Matthews et al. (2016), Amy Miller and colleagues attempted to quantify behavior and feeding and drinking habits of pigs to provide an automated system to indicate illness in pigs without the need for human observation. They found that while habits of the pigs can be analyzed with automation, social behavior must be observed by a human. They presented the challenge of developing a system that can detect social behavior. In Pastell et al. (2013), a study used PLF to predict farrowing (giving birth) in pigs in order to increase animal welfare.

In Chap. 9 of this book, Jarissa Maselyne (Flanders Research Institute for Agriculture, Fisheries, and Food, Belgium) discusses flagship EU projects in PLF that she has been involved in and her experience combining sensors, radio frequency identification (RFID), data analysis, decision support tools, Internet of Things (IoT), and High Performance Computing (HPC) to address some of the most pressing issues in pig farming. In particular, her aim is to use these innovative technologies to work toward more individualized tailor-made care for each pig. Her and similar research and development efforts will bring the sector from a mainly group-based management to individual monitoring in order to detect and maintain animal welfare.

1.3.3 Poultry

For poultry farmers, performing routine checks to monitor the health of poultry is a highly demanding and complex task (Sassi et al. 2016). Additionally, stress due to environmental factors is a common issue in the poultry industry. There's a great deal of research to apply PLF techniques to improve poultry production. Most of the studies in the literature deal with monitoring of the animal's welfare and methods to combat stress. PLF applied to the poultry industry can answer the public's concern for the welfare and health of the chickens (Sassi et al. 2016). A detailed review of studies using biotelemetry to monitor poultry physiological behavior under various environmental stressors is presented in Hamrita and Paulishen (2011). In Aydin et al. (2013), a study used a fully-automated monitoring system to analyze pecking sounds of broiler chickens. A correlation between pecking sounds and feeding habits was found, indicating that this system could analyze feed uptake in a noninvasive way to the chickens. As feed intake is monitored, farmers are aware of trends in eating and needs for more feed to ensure that chickens are being properly fed. In Fontana et al. (2015), a study used sensors to collect and analyze the vocalization of chickens in order to connect them to the growth of the birds. This study found a strong correlation between the frequencies of vocalizations and the weight of broilers across all production cycles. This research is the first step in creating a tool that can automatically track chicken growth rates based on their vocalizations.

In Chap. 8 of this book, Takoi Hamrita and her students Taylor Ogle and Amanda Yi (University of Georgia, USA) present highlights of the significant milestones achieved by her research program toward using biotelemetry to build precision poultry environmental controllers that respond directly and in real time to the needs of the birds. The discussion is presented in three sections: the first section focuses on biotelemetry and its use to monitor poultry deep body temperature (DBT) responses to various environmental conditions, the second section deals with DBT modeling efforts to date, and the third section presents results of the first poultry environmental controller prototype which responds to poultry DBT responses in real time.

1.4   Has Precision Agriculture Been Effective?

With an ever-increasing population pressure throughout the world and the need for increased agricultural production, there is a definite need for improved management of the world's agricultural resources (Díez et al. 2014). This need is being addressed by cutting-edge PA research and commercialization of resulting products and techniques on farms around the world. According to a study titled "Threats to Precision Agriculture," since the 2000s, PA has gained popularity among farmers, and there has been a great increase in using it (Boghossian et al. 2018). Another study analyzing the "progress and present standing of precision agriculture" concluded that PA provides the desired "production and environmental benefits" (Kapurkar et al. 2013). So how widely used has PA been, how effective has it been at delivering the promised results, and what are some of the challenges to its implementation?

1.4.1   Implementation of Precision Agriculture and Related Challenges

A study conducted by Margit Paustian and colleague analyzed how farm characteristics and farmer demographics affected the use of PA in German crop farmers (Paustian and Theuvsen 2016). They found that having additional farming businesses, having less than 5 years or between 16 and 20 years of experience, and having over 500 ha of land were positive factors in adopting PA. They also found that having less than 100 ha and growing barley were negative factors in using PA practices. From this research, it is evident that amount of experience, land area, and type of crop all influence the likelihood of using PA.

A study in Schimmelpfennig (2011) indicated that over 40% of farmers worldwide have adopted methods for yield monitoring. The authors indicate that producers using yield mapping had statistically higher yields. A study conducted by Michele Marra, Sherry Larkin, Jeanne Reeves, and colleagues investigated the influence of spatial yield variability on the number of precision farming technologies adopted, using a count data estimation procedure and farm-level data (Paxton et al. 2010). Results indicated that farmers with more within-field yield variability adopted a larger number of PA technologies. Additionally, younger and better educated producers and the number of PA technologies were significantly correlated. Finally, farmers using computers for management decisions also adopted a larger number of PA technologies.

A study in Yost et al. (2016) described the results of a PA system (PAS) being implemented on a 36 ha field in central Missouri, after a decade of yield and soil mapping and water quality testing. The system was implemented from 2004 to 2014. Studies showed that the system led to reduced temporal yield variation in the field. Spatial yield variation was not affected by the PAS. The study concluded that the greatest long-term advantage of a PAS was reduced temporal yield variation.

A challenge that reduces the adoption of PA is that farmers are unwilling to dedicate managerial time to analyzing data. The proposed solution is to outsource the data analysis and development of recommendations for farmers (Yousefi and Razdari 2015). Additionally, there is not enough research to prove the value of a pool of farm data, although indirect evidence shows that farm data has economic value (Coble et al. 2018). Also, there is still a need for federal laws to protect farm data. Until policymakers provide protection, farmers must trust who they share their data with. Currently, universities have partnered with farms to collect data for research. Existing transfer systems for this data are time-consuming and inefficient, and both parties would benefit from an improved system of transferring data, preferably wirelessly and in real time (Coble et al. 2018).

A study in Lima et al. (2018) assessed the use of PLF, in the form of electronic identification (EID) in Welsh and English farms. They found that farmers beliefs, demonstrated in the factors of external pressure, negative feelings, and practicality, affected their willingness to use the technology. This article suggests that better communication to farmers of the benefits of this technology and how to use it productively will increase the use of EID. Another article about the French sheep industry (Villeneuve et al. 2019) indicated that while sensors are available to gather data on the sheep, there is a need for data analysis and decision-making support for the farmer in order for PLF to be successful.

In order to use wireless technologies, such as wireless sensor networks, producers will need access to the Internet. Rural areas have a disadvantage because these areas do not have strong, reliable, and fast connection to broadband (Coble et al. 2018). Despite discrepancies in the adoption of PA, research by Dr. Lynn V. Dicks has revealed practices that farmers would like to implement even though they are not currently using them. These practices include prediction of pest and disease outbreaks, especially for livestock farmers and staff training on environmental issues (Dicks et al. 2018).

1.4.2   Economic, Environmental, and Consumer Benefits

An essential factor for the success and adoption of PA is its economic feasibility. Highly technological advancements such as variable rate input technologies struggle in some areas to receive the consideration and acceptance of farmers due to the lack of a conclusive amount of data, indicating that it is economically smart for a farmer to switch to these increasingly expensive methods (Schimmelpfennig 2011). Another obstacle farmers face is that decisions to switch to predominantly PA methods present a high-impact risk (Lasley 1998). An entire season's crop could be lost due to human error in misinterpretation of data maps, sensor programing, or variable rate sprayers (Lasley 1998). So the adoption of these technology information breakthroughs is extremely slow due to the caution and farm security, but as more data becomes available to the public, more technology will be adopted revealing both positive and negative impacts on agriculture (Lasley 1998).

A study conducted by Cláudia Silva, Márcia Azanha Moraes, and colleagues in Sao Paulo, Brazil, revealed a clear benefit with the adoption of PA technologies including increased yields, increased crop quality, and decreased costs (Silva et al. 2010). Another study conducted by Cláudia Silva, Sônia Vale, and colleagues in Mato Grosso do Sul, Brazil, served to mitigate fears of economic risks associated with PA. The study indicated that profitability indicators such as gross revenue and profitability indices have economic benefits correlating to precision systems and further assure investors of the reliable market that is precision farming (Silva et al. 2007). Overall, farmers are cautiously adopting technologies on the promise of fine-tuning their systems to bring home larger profits resulting from a streamlined agricultural process.

Along with economic benefits, PA has the potential to provide environmental benefits. Decisions informed by data collected through technology such as sensors and GIS can decrease waste of water and fertilizers. Agricultural runoff has been an environmental concern for decades. This runoff can contain chemicals that harm aquatic life if it flows into a body of water, yet farms are not regulated under the Clean Water Act (Coble et al. 2018). There are numerous field studies supporting this positive relationship between the environment and PA. In particular, Monica Díez found that water management is positively affected by the availability of reliable and fast geo-information (Díez et al. 2014). Additionally, splitting up the control of the area into sections and allocating resources automatically with PA technologies provided the greatest environmental enhancement with an improvement of 2.42% over the model without PA technologies (Brown et al. 2015). A field study was conducted at four sites in Denmark considering the change in fuel, pesticides, and herbicides usage when precision farming was implemented. The study specifically focused on the implementation of precision weed management and auto guidance equipment to cut down on inefficient fuel and pesticide spending and increase yield. The study showed that the adoption of PA technologies and techniques for the four main crops in Denmark led to increased income to farmers, decreased fuel consumption, and decreased use of pesticides/herbicides (Jensen et al. 2012). In Brazil, a countrywide survey revealing the adoption and benefits of PA technologies indicated higher yields, lower costs, minimization of environmental impacts, and improvements in sugarcane quality (Silva et al. 2010).

In a study conducted by Rachel Brown and colleagues, the environmental and economic effects stemming from the implementation of PA technologies were quantified for a corn and soybean farm based in Kentucky (Brown et al. 2015). The results indicated a 0.56% increase in the farm's economic mean net return and an improvement in the farm's environmental impact of 2.42%.

Economic benefit is not always evident in the use of PA. For example, numerous field studies have indicated that spatial variability includes scenarios where one parcel of land may need fewer inputs to thrive, and other areas may need more (Havránková et al. 2007). This property of spatially based input control occasionally balances out or increases the amount required by a site. One specific example is outlined controlling and monitoring the rate of nitrogen at sites in Slovakia and the United Kingdom (UK). Compared to the UK, which saw a reduction of 15 kg N/ha, there was no significant overall reduction in the total nitrogen used, and the overall cost of production using the sensors increased by 5% in Slovakia (Havránková et al. 2007).

Along with economic and environmental benefits, PA has the potential to provide consumer benefits. For instance, data management and data analytics can be used to provide consumers with information about the production of their food and compliance with production practice requirements (Coble et al. 2018).

1.5   What Educational Transformations Do We Need to Make to Keep Up with the Agricultural Revolution Created by Precision Agriculture?

PA involves the use of many cutting-edge technologies and tools such as GPS, GIS, yield monitoring systems, smart wireless sensing systems, sophisticated software platforms, vehicle guidance systems, and variable rate application equipment. As these technologies become more advanced, and as the benefits of PA become more apparent and it gains more popularity among farmers, it is becoming increasingly important that educational programs keep up with this progress.

Multiple studies have pointed out various educational gaps that are hindering progress in implementing PA. In Lamb et al. (2008), the authors highlight a large knowledge gap between developers of PA technology and users of it. The authors emphasize the need for universities to provide and promote PA topics and materials to college students, as well as the need for extension agents to educate producers. A study in Coble et al. (2018) highlighted the particular need for people with education and training in geospatial analysis and analytical techniques. A study in Kapurkar et al. (2013) indicated that there remains a gap between the knowledge of PA and the use of it. The reason for this gap is that developers are propelling the technology without farmers requesting it. If farmers are educated about the benefits of PA and taught how to use it, then this gap will diminish and PA might become more common practice. In Silva et al. (2007), Cláudia Silva and Sônia Vale attribute the reluctance to accept and implement PA to limited access to education and training. They indicate that in order for PA to provide the benefits to stakeholders (producers, agribusiness, and educators), they must improve their "agronomic knowledge and skills," "computer and information management skills," and "understanding of PA as a system for increasing knowledge." In Kitchen et al. (2002), Kitchen asserts that reluctance to implement PA seems to be based on accessibility to well-trained, knowledgeable people and the cost and availability to obtain quality education, training, and products. Kitchen adds that implementation of PA depends on the availability of education on PA and expanding access to all who are interested. Based on a study that surveyed Nebraskan-based agricultural producers at Extension Crop Production Clinics across the state, Extension Precision Ag Data Management Workshops, the 2015 Fremont Corn Expo (sponsored by Extension), and the 2015 NEATA Ag Technology Conference, there is an overall need for education and training regarding the adoption of PA technologies such as yield monitors, maps, GPS guidance, and the associated data (Castle et al. 2015). Agricultural producers must understand the issues that need to be addressed by precision farming techniques in order to make the most from the adoption of precision farming technologies.

There are several efforts to increase PA knowledge among the various stakeholders, and various research studies in the literature have examined these efforts and proposed a number of ways current educational systems need to evolve to support PA. In Skouby (2017), Danielle Skouby conducted a content review of 56 PA courses involving 44 institutions in the United States. The goal was to develop knowledge of what PA content is currently taught across the United States to help build a better understanding for what PA instructors should incorporate into their classes in the future. Information regarding topics taught in each course and time spent on each topic was gathered from each institution and analyzed. She found that topics most taught among universities were GPS, GIS, scope of PA, yield monitoring, and yield maps. She also determined that topics most discussed were practical, "how-to" topics, while topics not taught have more to do with environmental implications of PA. She concluded that educational programs could benefit from including more instruction beyond "how-to" do PA successfully to include instruction on environmental benefits and other implications of PA. Also, because of the variety among PA programs, Skouby also recognized a "need for a more standardized curriculum" so that students are best prepared to enter the industry (Skouby 2017).

In Shobha et al. (2008), Sriharan Shobha and colleagues reported on a capacity-building effort at Virginia State University (VSU) to incorporate PA into the curriculum. Since site-specific management is becoming heavily dependent on geographic information systems (GIS), the author and primary investigator (PI) worked with VSU and funding from the USDA Capacity Building Grant to introduce faculty to software packages such as ArcGIS for GIS and ERDAS for remote sensing. Then faculty members along with the PI incorporated GIS curricula into the Department of Agriculture and Human Ecology at VSU (Shobha et al. 2008).

In North Dakota State University (2018), the authors highlighted the importance of introducing new PA degrees into universities to ensure students are best equipped to join this field. Julie Bietz, student coordinator for the Department of Agricultural and Biosystems Engineering at North Dakota State University, states that "for our students to remain vital to the field, they must be educated in the newest technologies, methods, and practices for farming" (North Dakota State University 2018). North Dakota State University offers a major in PA that "teaches students how to manage, analyze, and use large amounts of digital data to increase production, profit, and better protect the environment" (North Dakota State University 2018).

In Heiniger et al. (2002), the authors propose that field days would be an ideal method of instruction that should be implemented in college-level PA programs to provide students with hands-on experiences and give them the opportunity to see the wide variety of "radically new technologies" and methods in agriculture. Field days and tours have traditionally been used to "introduce growers and agricultural professionals to new technologies and techniques" to see how they are applied (Heiniger et al. 2002). The article explains the benefits of field days and hands-on experiences in demonstrating the technology and techniques of PA to students.

In Kitchen et al. (2002), Kitchen indicates that by offering classes at workshops or conferences in addition to college-level programs, a greater demographic of students can learn beyond the traditional college student. He also proposes that as PA evolves, there will need to be different types and levels of classes that cater to the beginner or advanced precision farmer. Beginner students can be introduced to the basics, while more advanced students who understand the foundations can continue to stay on the cutting edge of new discoveries within PA (Kitchen et al. 2002).

1.6   Conclusion

PA relies heavily on technological and engineering innovations and extensive data collection. With mass amounts of data now readily available and the ability to capitalize on this digitized information, agriculture has officially entered the information age, and technology has become an integral part of the agricultural process. The ability to mine massive amounts of spatial and temporal farming data has led to innovations and breakthroughs such as GIS, variable rate technology, and robotics. Thanks to PA, the potential to manage farms on a microscale is now possible. The implementation of PA on farms has increased in the past few years (Coble et al.

2018). Farmers are seeing fewer inputs yield greater outputs due to the precise methods of production they have been able to adopt. In more tangible terms, farmers can use less water, less fertilizer, and less fuel in maintaining their crops. Tractors run more efficiently through autonomous control that removes human error. Variable rate sprayers coupled with spatial maps apply the precise amount of water and fertilizer for any given area. Not only is this amount precise, but it is precisely what that spatial area needs to yield quality crops. These developing tools and technologies serve as cost reducers, crop enhancers, and yield increasers. They also better address the issue of feeding an ever-growing population at a globally lower cost to farmers, the environment, and the consumer. With PA, the old way of life, agriculture, intersects with the new way of life, technology.

Getting the technology to be implemented on any kind of farm can still be challenging. The challenging versatility of PA can make it hard to convince producers to adopt PA technology and techniques (Bullock et al. 2007). Research has shown that young and informed producers are more likely to adopt PA technology (Paxton et al. 2010). For crop farmers, the likelihood of PA adoption increases for those with 16 or more years of experience or less than 5 years of experience (Paustian and Theuvsen 2016). Despite its name, PA is still improving in precision.

Without a doubt, PA is a multidisciplinary field that requires the collaboration of various stakeholders. In Bullock et al. (2007), the authors discuss the multidisciplinary facets of PA and present recommendations for creating multidisciplinary PA research. Precision technology from different industries, such as medical devices and appliances, can be applied to agriculture. As a result, when those industries change and advance, agriculture will be able to change and advance (Halachmi 2015). There is still much more work to be accomplished in research and development if we want to be prepared for a growing population.

Research in PA has immense potential. PA has shown promise to improve and increase yield and conserve resources. Growth in research is expected due to promotion of STEM fields in schools and the global and urgent need for innovation in agriculture. More diverse research is still needed for long-term impacts of PA practices (Yost et al. 2016). Overall, technology has catapulted agricultural development despite challenges. Research has proven that technology – GPS, GIS, remote sensing, smart sensors, variable rate technology, nanotechnology, RFID, IoT, yield monitors, and UAVs – is vital for agriculture. We can expect that PA will become more and more relevant over the upcoming years. When farmers and researchers can work together and when people with desk jobs collaborate with people with field jobs, food production issues can be solved.

PA is a natural fit for women. Our land, our environment, and natural resources are stressed to the max. New solutions to steer us in a more positive and sustainable direction are becoming more vital every day. Women are driven by the common good, they like pursuing passions, and they care about their communities, the environment, animals, etc. PA is inherently designed to optimize the use of natural resources and reduce waste, and to contribute to our well-being and that of the environment and the animals within it. A natural fit. Women are a vital intellectual, emotional, and spiritual resource for vital new global paradigms such as PA.

References

Adamchuk, V. I., Morgan, M. T., & Ess, D. R. (1999). An automated sampling system for measuring soil pH. *Transactions of ASAE, 42*, 885–892. https://doi.org/10.13031/2013.13268.

Adrion, F., Kapun, A., Eckert, F., Holland, E.-M., Staiger, M., Götz, S., & Gallmann, E. (2018). Monitoring trough visits of growing-finishing pigs with UHF-RFID. *Computers and Electronics in Agriculture, 144*, 144–153. https://doi-org.proxy-remote.galib.uga.edu/10.1016/j.compag.2017.11.036.

Aguilar-Rivera, N., Algara-Siller, M., Olvera-Vargas, L. A., & Michel-Cuello, C. (2018). Land management in Mexican sugarcane crop fields. *Land Use Policy, 78*, 763–780. https://doi.org/10.1016/j.landusepol.2018.07.034.

Alsaaod, M., Schaefer, A., Büscher, W., & Steiner, A. (2015). The role of infrared thermography as a non-invasive tool for the detection of lameness in cattle. *Sensors, 15*, 14513–14525. https://doi.org/10.3390/s150614513.

Arslan, S., & Colvin, T. S. (2002). Grain yield mapping: Yield sensing, yield reconstruction, and errors. *Precision Agriculture, 3*, 135–154.

Aydin, A., Bahr, C., & Berckmans, D. (2013). *An innovative monitoring system to measure the feed intake of broiler chickens using pecking sounds.* Precision livestock farming 2013 – Papers presented at the 6th European Conference on Precision Livestock Farming, ECPLF 2013, pp. 926–936.

Banhazi, T. M., Babinszky, L., Halas, V., & Tscharke, M. (2012). Precision livestock farming: Precision feeding technologies and sustainable livestock production. *International Journal of Agricultural & Biological Engineering, 5*(4), 54–61. https://doi-org.proxy-remote.galib.uga.edu/10.3965/j.ijabe.20120504.006.

Baptista, E. S., Baptista, F. J., & Castro, J. A. (2013). *Environmental and endocrine assessment of sheep welfare in a climate-controlled room.* 6th European conference on precision livestock farming, pp. 397–406.

Boghossian, A., Linsky, S., & Brown, A. (2018). *Threats to precision agriculture.* 2018 Public-private analytic exchange program.

Brown, R. M., Dillon, C. R., Schieffer, J., & Shockley, J. M. (2015). The carbon footprint and economic impact of precision agriculture technology on a corn and soybean farm. *Journal of Environmental Economics and Policy, 5*, 335–348. https://doi.org/10.1080/21606544.2015.1090932.

Bullock, D. S., Kitchen, N., & Bullock, D. G. (2007). Multidisciplinary teams: A necessity for research in precision agriculture systems. *Crop Science, 47*(5), 1765–1769. https://doi.org/10.2135/cropsci2007.05.0280

Cameron, K., & Hunter, P. (2002). Using spatial models and kriging techniques to optimize long-term ground-water monitoring networks: A case study. *Environmetrics, 13*, 629–656. https://doi.org/10.1002/env.582.

Caria, M., Sara, G., Todde, G., Polese, M., & Pazzona, A. (2019). Exploring smart glasses for augmented reality: A valuable and integrative tool in precision livestock farming. *Animals, 9*(11), pii: E903. https://doi-org.proxy-remote.galib.uga.edu/10.3390/ani9110903.

Castle, M., Lubben, B & Luck, J. (2015). *Precision agriculture usage and big agriculture data.* https://agecon.unl.edu/cornhusker-economics/2015/precision-agriculture-usage-and-big-agriculture-data. Accessed 21 June 2019.

Coble, K. H., Mishra, A., Ferrell, S., & Griffin, T. (2018). Big data in agriculture: A challenge for the future. *Applied Economic Perspectives and Policy, 40*, 79–96.

D'Eath, R. B., Jack, M., & Futro, A. (2018). Automatic early warning of tail biting in pigs: 3D cameras can detect lowered tail posture before an outbreak. *PLoS One, 13*(4), e0194524. https://doi.org/10.1371/journal.pone.0194524.

Dalezios, N. R., Dercas, N., Spyropoulos, N. V., & Psomiadis, E. (2017). Water availability and requirements for precision agriculture in vulnerable agroecosystems. *European Water, 59*, 387–394.

Dalezios, N. R., Dercas, N., Spyropoulos, N. V., & Psomiadis, E. (2019). Remotely sensed methodologies for crop water availability and requirements in precision farming of vulnerable agriculture. *Water Resources Management, 33*, 1499–1519. https://doi.org/10.1007/s11269-018-2161-8.

Dicks, L. V., Rose, D. C., & Ang, F. (2018). What agricultural practices are most likely to deliver 'sustainable intensification' in the UK? *Food and Energy Security, 8*(1), e00148. https://doi.org/10.1002/fes3.148.

Díez, M., Moclan, C., Romo, A., & Pirondini, F. (2014). *High-resolution super-multitemporal monitoring: Two-day time series for precision agriculture applications.* https://iafastro.directory/iac/archive/browse/IAC-14/B1/5/26742/

Duhan, J. S., Kumar, R., & Kumar, N. (2017). Nanotechnology: The new perspective in precision agriculture. *Biotechnology Reports, 15*, 11–23. https://doi.org/10.1016/j.btre.2017.03.002.

Elarab, M., Ticlavilca, A. M., & Torres-Rua, A. F. (2015). Estimating chlorophyll with thermal and broadband multispectral high resolution imagery from an unmanned aerial system using relevance vector machines for precision agriculture. *International Journal of Applied Earth Observation and Geoinformation, 43*, 32–42. https://doi.org/10.1016/j.jag.2015.03.017.

Fontana, I., Tullo, E., & Fernandez, A. (2015). *Frequency analysis of vocalisation in relation to growth in broiler chicken.* 7th European conference on precision livestock farming, At Milan, Italy.

Fort, H., Dieguez, F., Halty, V., & Lima, J. M. S. (2017). Two examples of application of ecological modeling to agricultural production: Extensive livestock farming and overyielding in grassland mixtures. *Ecological Modelling, 357*, 23–34. https://doi.org/10.1016/j.ecolmodel.2017.03.023.

Gago, J., Douthe, C., & Coopman, R. (2015). UAVs challenge to assess water stress for sustainable agriculture. *Agricultural Water Management, 153*, 9–19. https://doi.org/10.1016/j.agwat.2015.01.020.

Gómez-Candón, D., Castro, A. I. D., & López-Granados, F. (2013). Assessing the accuracy of mosaics from unmanned aerial vehicle (UAV) imagery for precision agriculture purposes in wheat. *Precision Agriculture, 15*, 44–56. https://doi.org/10.1007/s11119-013-9335-4.

Haboudane, D., Miller, J. R., & Tremblay, N. (2002). Integrated narrow-band vegetation indices for prediction of crop chlorophyll content for application to precision agriculture. *Remote Sensing of Environment, 81*, 416–426. https://doi.org/10.1016/s0034-4257(02)00018-4.

Halachmi, I. (2015). *Precision livestock farming applications: Making sense of sensors to support farm management.* Wageningen: Wageningen Academic.

Hamrita, T., & Paulishen, M. (2011). Advances in management of poultry production using biotelemetry. In *Modern telemetry.* Rijeka, Croatia: InTech. https://doi.org/10.5772/24691.

Hamrita, T., Tollner, E., & Schafer, R. (1996). *Towards a robotic farming vision: Advances in sensors and controllers for agricultural system applications.* IAS 96 conference record of the 1996 IEEE industry applications conference thirty-first IAS annual meeting. https://doi.org/10.1109/ias.1996.559293.

Hamrita, T., Tollner, E., & Schafer, R. (2000). Toward fulfilling the robotic farming vision: Advances in sensors and controllers for agricultural applications. *IEEE Transactions on Industry Applications, 36*, 1026–1032. https://doi.org/10.1109/28.855956.

Hamrita, T., Kaluskar, N., & Wolfe, K. (2005). *Advances in smart sensor technology.* Fortieth IAS annual meeting conference record of the 2005 industry applications conference. https://doi.org/10.1109/ias.2005.1518731.

Havránková, J., Godwin, R. J., & Wood, G. A. (2007). *The evaluation of ground based remote sensing systems for canopy nitrogen management in winter wheat.* Silsoe: Cranfield University.

Heiniger, R. W., Havlin, J. L., Crouse, D. A., & Knowles, T. (2002). Seeing is believing: The role of field days and tours in precision agriculture education. *Precision Agriculture, 3*, 309–318.

Important tools to succeed in precision farming. (2018). *Precision agriculture.* https://precisionagricultu.re/important-tools-to-succeed-in-precision-farming/. Accessed 20 Mar 2020.

Ivanov, S., Bhargava, K., & Donnelly, W. (2015). Precision farming: Sensor analytics. *IEEE Intelligent Systems, 30*, 76–80. https://doi.org/10.1109/mis.2015.67.

Jensen H. G., Jacobsen L. B., Pedersen S. M., & Tavella E. (2012). Socioeconomic impact of widespread adoption of precision farming and controlled traffic systems in Denmark. *Precision Agriculture, 13*(6):661–677.

Jørgensen, R. N., Sørensen, C. G., & Jensen, H. F. (2007, June 17–20). *FeederAnt – An autonomous mobile unit feeding outdoor pigs*. 2007 Minneapolis, Minnesota. https://doi.org/10.13031/2013.22864.

Kah, M., & Hofmann, T. (2014). Nanopesticide research: Current trends and future priorities. *Environment International, 63*, 224–235. https://doi.org/10.1016/j.envint.2013.11.015.

Kapurkar, P. M., Kurchania, A. K., & Kharpude, S. N. (2013). GPS and remote sensing adoption in precision agriculture. *International Journal of Agricultural Engineering, 6*, 221–226.

Kerry, R., Oliver, M. A., & Frogbrook, Z. L. (2010). Sampling in precision agriculture. In *Geostatistical applications for precision agriculture* (pp. 35–63). Dordrecht: Springer. https://doi.org/10.1007/978-90-481-9133-8_2.

Kitchen, N. R., Snyder, C. J., Franzen, D. W., & Wiebold, W. J. (2002). Educational needs of precision agriculture. *Precision Agriculture, 3*, 341–351.

Kühn, J., Brenning, A., & Wehrhan, M. (2008). Interpretation of electrical conductivity patterns by soil properties and geological maps for precision agriculture. *Precision Agriculture, 10*, 490–507. https://doi.org/10.1007/s11119-008-9103-z.

Kutz, L. J., Miles, G. E., Hammer, P. A., & Krutz, G. W. (1987). Robotic transplanting of bedding plants. *Transactions of ASAE, 30*, 0586–0590. https://doi.org/10.13031/2013.30443.

Lamb, D. W., Frazier, P., & Adams, P. (2008). Improving pathways to adoption: Putting the right Ps in precision agriculture. *Computers and Electronics in Agriculture, 61*, 4–9. https://doi.org/10.1016/j.compag.2007.04.009.

Lasley, P. (1998). *Perceived risks and decisions to adopt precision farming methods*. 4c Precision Ag Edition 9–9.

Lima, E., Hopkins, T., & Gurney, E. (2018). Drivers for precision livestock technology adoption: A study of factors associated with adoption of electronic identification technology by commercial sheep farmers in England and Wales. *PLoS One, 13*(1), e0190489. https://doi.org/10.1371/journal.pone.0190489.

Ma, Y., Wu, H., & Wang, L. (2014). Remote sensing big data computing: Challenges and opportunities. *Future Generation Computer Systems, 51*, 47–60. https://doi.org/10.1016/j.future.2014.10.029.

Marek, T., Almas, L., Amosson, S., & Cox, E. (2001). *The feasibility of variable rate irrigation with center pivot systems in the Northern Texas High Plains*. 2001 Sacramento, CA, 29 July–1 August 2001. https://doi.org/10.13031/2013.3443.

Martelloni, L., Fontanelli, M., Frasconi, C., et al. (2016). Cross-flaming application for intra-row weed control in maize. *Applied Engineering in Agriculture, 32*, 569–578. https://doi.org/10.13031/aea.32.11114.

Maselyne, J., Saeys, W., & Nuffel, A. V. (2013). *A health monitoring system for growing-finishing pigs based on the individual feeding pattern using radio frequency identification and synergistic control*. Papers presented at the 6th European conference on precision livestock farming, Leuven, pp. 825–833.

Matthews, S. G., Miller, A. L., & Clapp, J. (2016). Early detection of health and welfare compromises through automated detection of behavioural changes in pigs. *The Veterinary Journal, 217*, 43–51. https://doi.org/10.1016/j.tvjl.2016.09.005.

Morris, J. E., Cronin, G. M., & Bush, R. D. (2012). Improving sheep production and welfare in extensive systems through precision sheep management. *Animal Production Science, 52*, 665. https://doi.org/10.1071/an11097.

Mulla, D., & Khosla, R. (2016). Historical evolution and recent advances in precision farming. In R. Lal & B. A. Stewart (Eds.), *Soil-specific farming precision agriculture* (pp. 1–35). Boca Raton: CRC Press. https://doi.org/10.1201/b18759-2.

National Museum of American History. (2018). Precision farming. In *National Museum of American History*. Smithsonian. https://americanhistory.si.edu/american-enterprise-exhibition/new-perspectives/precision-farming. Accessed 20 Mar 2020.

North Dakota State University. (2018). *New precision ag major offered at NDSU – College of Agriculture, Food Systems, and Natural Resources*. https://www.ag.ndsu.edu/academics/new-precision-ag-major-offered-at-ndsu. Accessed 27 Jan 2020.

O'Shaughnessy, S. A., Evett, S. R., & Colaizzi, P. D. (2019). Identifying advantages and disadvantages of variable rate irrigation: An updated review. *Applied Engineering in Agriculture, 35*, 837–852. https://doi.org/10.13031/aea.13128.

Ozguven, M. M. (2018). The newest agricultural technologies. *Current Investigations in Agriculture and Current Research, 5*(1), 573–580. https://doi.org/10.32474/ciacr.2018.05.000201.

Pandey, G. (2018). Challenges and future prospects of agri-nanotechnology for sustainable agriculture in India. *Environmental Technology and Innovation, 11*, 299–307. https://doi.org/10.1016/j.eti.2018.06.012.

Pastell, M., Hietaoja, J., & Yun, J. (2013). *Predicting farrowing based on accelerometer data*. The 6th European Conference on Precision Livestock Farming (EC-PLF 2013), Leuven, Belgium.

Paustian, M., & Theuvsen, L. (2016). Adoption of precision agriculture technologies by German crop farmers. *Precision Agriculture, 18*, 701–716. https://doi.org/10.1007/s11119-016-9482-5.

Paxton, K. W., Mishra, A. K., & Chintawar, S. (2010, February 6–9). *Precision agriculture technology adoption for cotton production*. Selected Paper prepared for presentation at the Southern Agricultural Economics Association annual meeting, Orlando, FL.

Peña, J. M., Torres-Sánchez, J., & Castro, A. I. D. (2013). Weed mapping in early-season maize fields using object-based analysis of unmanned aerial vehicle (UAV) images. *PLoS One, 8*(10), e77151. https://doi.org/10.1371/journal.pone.0077151.

Phadikar, S., Das, A. K., & Sil, J. (2012, January). Misclassification and cluster validation techniques for feature selection of diseased rice plant images. *Advances in intelligent and soft computing proceedings of the international conference on information systems design and intelligent applications 2012 (INDIA 2012)* held in Visakhapatnam, India, pp. 137–144. https://doi.org/10.1007/978-3-642-27443-5_16.

Poulopoulou, I., & Chatzipapadopoulos, F. (2015). *Saving resources using a cloud livestock farm management tool*. 7th European conference on precision livestock farming, pp. 276–283.

Pudumalar, S., Ramanujam, E., & Rajashree, R. H. (2017). *Crop recommendation system for precision agriculture*. 2016 eighth International Conference on Advanced Computing (ICoAC). https://doi.org/10.1109/icoac.2017.7951740.

Richard, M.-M., Sloth, K. H., & Veissier, I. (2015). *Real time positioning to detect early signs of welfare problems in cows*. European conference on precision livestock farming, Milan, Italy, 4pp.

Rodríguez, S., Gualotuña, T., & Grilo, C. (2017). A system for the monitoring and predicting of data in precision agriculture in a rose greenhouse based on wireless sensor networks. *Procedia Computer Science, 121*, 306–313. https://doi.org/10.1016/j.procs.2017.11.042.

Rovira-Más, F., Millot, C., & Sáiz-Rubio, V. (2015). *Navigation strategies for a Vineyard Robot*. 2015 ASABE international meeting. https://doi.org/10.13031/aim.20152189750.

Ruiz-Garcia, L., Lunadei, L., Barreiro, P., & Robla, I. (2009). A review of wireless sensor technologies and applications in agriculture and food industry: State of the art and current trends. *Sensors, 9*, 4728–4750. https://doi.org/10.3390/s90604728.

Sáiz-Rubio, V. A., Rovira-Más, F. A., Broseta-Sancho, P. A., & Aguilera-Hernández, R. A. (2015). *Robot-generated crop maps for decision-making in Vineyards*. 2015 ASABE international meeting. https://doi.org/10.13031/aim.20152189909.

Sassi, N. B., Averós, X., & Estevez, I. (2016). Technology and poultry welfare. *Animals, 6*, 62. https://doi.org/10.3390/ani6100062.

Schimmelpfennig, D. (2011). *On the doorstep of the information age: Recent adoption of precision agriculture*. Washington, DC: U.S. Department of Agriculture, Economic Research Service.

Schrøder-Petersen, D. I., & Simonsen, H. B. (2001). Tail biting in pigs. *The Veterinary Journal, 162*, 196–210. https://doi.org/10.1053/tvjl.2001.0605.

Shobha, S., Everitt, J. H., & Fletcher, R. (2008). *Geographic information system (GIS) and remote sensing (RS): Undergraduate academic curriculum and precollege training program*. IGARSS 2008 – 2008 IEEE International Geoscience and Remote Sensing Symposium. https://doi.org/10.1109/igarss.2008.4779628.

Silva, C. B., Do Vale, S. M. L. R., & Pinto, F. A. C. (2007). The economic feasibility of preci-sion agriculture in Mato Grosso do Sul State, Brazil: A case study. *Precision Agriculture, 8*, 255–265. https://doi.org/10.1007/s11119-007-9040-2.

Silva, C. B., De Moraes, M. A. F. D., & Molin, J. P. (2010). Adoption and use of precision agricul-ture technologies in the sugarcane industry of São Paulo state, Brazil. *Precision Agriculture, 12*, 67–81. https://doi.org/10.1007/s11119-009-9155-8.

Skouby, D. (2017). *A content review of precision agriculture courses across the United States.* International conference on precision agriculture, July 31–August 4, 2016, St. Louis, Missouri. Available: https://www.ispag.org/proceedings/?action=abstract&id=2186

Srinivasan, A. (2006). *Handbook of precision agriculture.* New York: Food Products Press. https://doi.org/10.1201/9781482277968.

Tarrío, P. M., Bernardos, A. M., Casar, J. R., & Besada, J. A. (2006, July 23–25). *A harvesting robot for small fruit in bunches based on 3-D stereoscopic vision.* Computers in agriculture and natural resources, Orlando, Florida. https://doi.org/10.13031/2013.21885.

Technology Quarterly. (2016). The future of agriculture. *The Economist.* https://www.economist.com/technology-quarterly/2016-06-09/factory-fresh. Accessed 20 Mar 2020.

Thornton, P. K. (2010). Livestock production: Recent trends, future prospects. *Philosophical Transactions of the Royal Society, B: Biological Sciences, 365*, 2853–2867. https://doi.org/10.1098/rstb.2010.0134.

Turner, L. W., Udal, M. C., Larson, B. T., & Shearer, S. A. (2000). Monitoring cattle behavior and pasture use with GPS and GIS. *Canadian Journal of Animal Science, 80*, 405–413. https://doi.org/10.4141/a99-093.

Vasconez, J. P., Cantor, G. A., & Cheein, F. A. A. (2019). Human–robot interaction in agriculture: A survey and current challenges. *Biosystems Engineering, 179*, 35–48.

Vellidis, G., Perry, C. D., & Durrence, J. S. (2001). The peanut yield monitoring system. *Transactions of ASAE, 44*(4), 775–785.

Villeneuve, É., Akle, A. A., & Merlo, C. (2019). Decision support in precision sheep farming. *IFAC-Papers OnLine, 51*, 236–241. https://doi.org/10.1016/j.ifacol.2019.01.048.

Wishart, H., Morgan-Davies, C., & Waterhouse, A. (2015). A PLF approach for allocating supple-mentary feed to pregnant ewes in an extensive hill sheep system. *Precision Livestock Farming, 15*, 256–265.

Yost, M. A., Kitchen, N. R., & Sudduth, K. A. (2016). Long-term impact of a precision agriculture system on grain crop production. *Precision Agriculture, 18*, 823–842.

Yousefi, M. R., & Razdari, A. M. (2015). Application of Gis and Gps in precision agriculture (a review). *International Journal of Advanced Biological and Biomedical Research, 3*, 7–9.

Yu, P., Li, C., Rains, G., & Hamrita, T. (2011a). Development of the berry impact recording device sensing system: Hardware design and calibration. *Computers and Electronics in Agriculture, 79*, 103–111. https://doi.org/10.1016/j.compag.2011.08.013.

Yu, P., Li, C., Rains, G., & Hamrita, T. (2011b). Development of the berry impact recording device sensing system. *Software Computers and Electronics in Agriculture, 77*, 195–203. https://doi.org/10.1016/j.compag.2011.05.003.

Yun, G., Mazur, M., & Pederii, Y. (2017). Role of unmanned aerial vehicles in precision farming. *Proceedings of the National Aviation University, 1*(70), 106–112. https://doi.org/10.18372/2306-1472.70.11430.

Zehner, N., Niederhauser, J. J., Schick, M., & Umstatter, C. (2019). Development and validation of a predictive model for calving time based on sensor measurements of ingestive behavior in dairy cows. *Computers and Electronics in Agriculture, 161*, 62–71. https://doi.org/10.1016/j.compag.2018.08.037.

Zhang, C., & Kovacs, J. M. (2012). The application of small unmanned aerial systems for pre-cision agriculture: A review. *Precision Agriculture, 13*, 693–712. https://doi.org/10.1007/s11119-012-9274-5.

Dr. Takoi Khemais Hamrita is a Professor of Electrical Engineering at the University of Georgia, where she has spearheaded the development of two ABET-accredited degree programs, one in electrical and the other in computer systems engineering. These efforts have recently culminated into a new UGA school of electrical and computer engineering for which she has served as inaugural chair. Dr. Hamrita has been at UGA for 25 years, where she was the first woman faculty member to be hired into the Department of Biological and Agricultural Engineering as an Assistant Professor (she remained the only female Professor in her department for almost 15 years). Dr. Hamrita's main research focus is on precision agriculture. She has worked in many areas of precision agriculture including yield monitoring, smart poultry environmental control, biotelemetry, smart sensing, and harvest and post-harvest technology. She holds a patent in yield monitoring, has published over 50 articles and book chapters, and has given over 50 conference presentations around the world on related research. Dr. Hamrita has served in many leadership roles within professional societies including:

- Chair, Vice Chair, and Secretary of the IEEE-IAS (Industry Applications Society) – Industrial Automation and Control Committee
- Associate Editor for the Industry Applications Society journals IEEE-IAS Conference technical program chair
- Southeast coordinator and IEEE USA liaison for Women in Engineering (WIE).
- Chair, Vice Chair, and Secretary of the ASAE (American Society of Agricultural Engineers)-IET (Instrumentation and Control Committee)
- Comparative and International Education Society (CIES) session organizer at the CIES Annual International Conference

American Society of Engineering Education (ASEE) session organizer at the ASEE Annual International Conference. Dr. Hamrita is a fierce advocate for women in Science, Technology, Engineering and Math (STEM) and is the founder and chair of the Global Women in STEM Leadership Summit. The program is an ongoing movement that aims to educate, inspire, empower and elevate women and girls in scientific and technological fields to help eliminate internal and external barriers they face. The women come from all career paths and stages from high school to the C-Suite. Our goal is to build capacity, nurture talent, create community, and empower women and girls in STEM to reach their full potential and thrive in male dominated fields. The summit convenes some of the most successful and influential leaders and founders from industry, academia, nonprofit and government. Dr. Hamrita has built a decade-long partnership between UGA and the Tunisian Ministry of Higher Education, which has had profound impact both on UGA and Tunisia and has become an innovative model for education and development around the world. Some of the most notable impacts of this partnership is the launching of a national virtual university in Tunisia that is currently delivering, online, a sizeable portion of higher education curricula across disciplines. The program has earned her numerous prestigious awards such as the Tunisian

National Medal in Science and Education, the Andrew Heiskell Award for Innovation in International Partnerships, and the Tunisian Community Center's Ibn Khldoun Award for Excellence in Public Service.

Kaelyn Deal is an undergraduate studying Mechanical Engineering at the University of Georgia. She is an active undergraduate researcher participating in the Center for Undergraduate Research Opportunities (CURO) and the Symposium on Space Innovations. In addition, she has worked at the UGA Small Satellite Research Laboratory for 2 years as a Mechanical Systems and Thermal Engineer.

Selyna Gant is an undergraduate student at the University of Georgia studying agricultural engineering. Her degree emphasis is electrical engineering. On campus, she is involved in the American Society of Agricultural and Biological Engineers and the Navigators. In the Athens community, she is a volunteer at the Athens Area Homeless Shelter. Her plans after graduation are to pursue a master's degree in agricultural engineering. Her research interest areas are embedded systems, sustainability, irrigation systems, and nutrient management.

Haley Selsor is a senior at the University of Georgia studying agricultural engineering. Her area of emphasis is natural resources, and she is interested in a career in water resources engineering after graduation. Outside of classes, Haley works as a peer tutor for UGA's Academic Resource Center and is heavily involved at her church in Athens, GA.

Chapter 2
Sensing Technologies and Automation for Precision Agriculture

Man Zhang, Ning Wang, and Liping Chen

Contents

2.1 Introduction

Agriculture production systems have benefited from the incorporation of technological advances primarily developed for other industries. The industrial age brought mechanization and synthesized fertilizers to agriculture. The technology age provided genetic engineering to crop breeding and automation to agricultural operations. Now the information age brings the potential to integrate data-centric information technologies to achieve site-specific agricultural production, i.e.,

M. Zhang
College of Information and Electrical Engineering, China Agricultural University, Beijing, China

N. Wang (✉)
Department of Biosystems and Agricultural Engineering, Oklahoma State University, Stillwater, OK, USA
e-mail: ning.wang@okstate.edu

L. Chen
Beijing Research Center of Intelligent Equipment for Agriculture, Beijing Academy of Agriculture and Forestry Sciences, Beijing, China

© Springer Nature Switzerland AG 2021
T. K. Hamrita (ed.), *Women in Precision Agriculture*, Women in Engineering and Science, https://doi.org/10.1007/978-3-030-49244-1_2

precision agriculture (PA). PA technology has been promoted and implemented around the world in the last decade. The factual base of PA is the spatial and temporal variability of soil and crop factors between and within fields. Before the introduction of agricultural mechanization, the very small size of fields allowed farmers to monitor field conditions and vary treatments manually. With the enlargement of fields and intensive mechanization, crops have been treated under "average/uniform" soil, nutrient, moisture, weed, insect, and growth condition assumptions. This has led to over-/under-applications of herbicides, pesticides, irrigation, and fertilizers.

PA is conceptualized by a system approach to reorganize the total system of agriculture toward a low-input, high-efficiency, and sustainable agriculture (Shibusawa 1998). This new approach benefits from the emergence and convergence of several technologies, including the Global Positioning System (GPS), geographic information system (GIS), miniaturized computer components, automatic control, in-field and remote sensing, mobile computing, advanced information processing, and telecommunications (Gibbons 2000). The agricultural industry is now capable of acquiring detailed "knowledge" on production variability, both spatially and temporally, and automatically adjusting treatments to meet each site's unique needs. Various sensors and actuators with intelligence (namely "smart sensors") have discovered a great arena in agricultural field data acquisition, monitoring, and control. Their major shortcomings are the requirements of extensive wiring for multiple-point measurement and control, frequent on-site data downloading, in-time maintenance due to unavoidable damages, and loss due to weather conditions and thefts. Failure to satisfy these requirements may lead to loss of data and malfunction of the overall system. PA has been further extended to livestock production, namely, precision ranching, which aims to improve the productivity and sustainability of rangelands by grazing *the right animal at the right place at the right time*. With the use of networked, miniaturized electronic devices, the animal behavior and health conditions, as well as forage conditions, can be monitored, recorded, and controlled in a more accurate way in order to optimize animal protein production and grassland management.

While the principles of PA have been developed for around 30 years, the adoption of PA technologies achieved a large leap only over the past 15 years due to rapid advancements and low-cost deployment of electronic, computing, Internet, mobile network, big data, and artificial intelligence technologies. Networked smart sensors and controllers are embedded in farm equipment to maximize their capability and optimize their productivity and efficiency. Stationary smart sensors have been developed and used to monitor agricultural environment, crop growth, soil conditions, and disease/insect damages. Crop scientists and farmers have quickly adopted drone-based/aerial-based sensing systems to collect various crop data over large-scale farmland. The goal of PA approach is to *improve the productivity and sustainability of agricultural production by treating the right plant at the right place at the right time*. With the advancement of smart sensors, agricultural production management can make relevant, specific decisions on crop/field treatments and machinery operations based on high-resolution and high-throughput in-field spatial and time information. In this chapter, we will provide a summary on the commonly used sensors in PA applications.

2.2 Crop Sensing Technology

Various sensors have been developed and used to monitor crop growth status. Plant height, canopy temperature, leaf chlorophyll content, leaf wetness, etc. are the most used indicators of health status of plants. Many commercial sensors are available on the market, while research is still going on to optimize the approaches of crop sensing. Some of the sensing techniques use contact measurements, which require the sensor(s) to be mounted on a leaf or a stem. Others can be installed on mobile platforms to conduct noncontact and on-the-go measurements.

Optical sensors are a big category in crop sensing techniques. They provide measurements of spectral reflectance of canopies at different wavebands, which are used to calculate normalized difference vegetation index (NDVI) values. The NDVI values are often used to determine the nutrient level, leaf chlorophyll content, leaf area index, etc. Three prominent commercial sensors in this category include GreenSeeker™ (http://www.trimble.com/agriculture), the Crop Circle™ (holland-scientific.com), and CropSpec™ (topcon.com). They are all based on real-time, on-the-go measurements of multiple optical sensors to determine crop growth status and make decisions on nitrogen applications. The primary differences between these sensors are the wavelengths that each uses for sensing. Skye Instruments Ltd. based in Britain provides a wide range of portable sensors and systems for plant and crop monitoring including plant moisture potential, leaf area, root length, etc. A French company, Force-A (www.force-a.eu), provides sensors to measure polyphenol and chlorophyll concentration in leaves based on plant's intrinsic fluorescence characteristics.

Many sensing methods were developed to measure crop height. Stereo vision is a method of distance measurement using the triangle calculation based on different perspectives of a scene. Its applications are mainly in distance measurement and automatic guidance. Rovira-Más et al. (2003) used a compact stereo camera mounted on an unmanned helicopter to map plant height in corn fields. Bulanon et al. (2004) used the stereo vision technology to locate the position of apples during harvesting. The major disadvantage of stereo vision technology is the expensive computation and low accuracy. An ultrasonic sensor measures the distance between the sensor and an object by sending out sound waves and receiving the bounced back echoes. Knowing the speed of sound waves under certain environmental temperature and the time of flight the waves traveled, the distance can be calculated. Because of the nature of sound waves, ultrasonic sensors have relatively wider beam angles. Hence, they are suitable for situations without critical spatial resolution requirements. Also, they are robust to environmental interference, such as sunlight, dust, and water compared with other distance sensors. Sui et al. (2012, 2013) utilized ultrasonic sensors for vehicle-based measurement coupled with a Global

Positioning System (GPS) to measure height of cotton plants and generate height distribution maps. Sharma et al. (2016) developed an on-the-go sensing system for variability mapping in a cotton field. They measured the cotton canopy height at different growth stages using ultrasonic sensors. A good correlation was obtained between the manually measured heights and sensor measured heights ($R^2 > 0.80$). Swain et al. (2009) used ultrasonic sensors to map a blueberry field in order to differentiate weed, wild blueberry, and bare spots. Jones et al. (2004) also used an ultrasonic sensor to measure the height of corn, spinach, and snap beans under lab conditions. The sensor was mounted directly above the plant and its beam angle was 12°. They indicated that the correlation between the sensor measured height and the manual measurement was $R^2 = 0.87$ for all three kinds of crop.

Light Detection and Ranging (LiDAR) sensors are based on another approach of distance sensing which senses the distance by detecting the time of flight or the phase shift between emitted and reflected lights. A laser emits a light in a narrow and low divergence beam, even after the light travels a long distance. This characteristic of laser gives it a great potential of high-resolution distance sensing. A near-infrared laser is a laser which has its light source in the near-infrared waveband. Airborne laser range finders have been widely used in photogrammetry and remote sensing for digital terrain modeling, large-scale crop surface mapping, crop shape profiling, and density estimation. Wei and Salyani (2004, 2005) designed and tested a LiDAR system to quantify foliage density of citrus trees. For ten citrus trees, the laser measurements had a good repeatability and correlated well with visual assessments ($R^2 = 0.96$). Saeys et al. (2009) estimated wheat stand density by measuring the variations in a laser penetration depth. Good crop density estimation was obtained ($R^2 = 0.80 \sim 0.96$). Shi et al. (2013, 2015) used LiDAR to successfully measure corn height, plant space, and population.

Range cameras provide real-time depth information in the form of an image. These cameras are mostly based on the time-of-flight principle and generate images with the amplitude of each pixel being the distance between the camera and the measured object. Many of these range cameras are called RGB-D cameras and are equipped with an RGB (red, green, and blue) camera, a depth imager, and an infrared emitter. They simultaneously provide color image, a depth image, and an infrared image of an object of interest. Kinect camera from Microsoft, Xtion from Asus, and RealSense from Intel are regarded as representatives of such cameras and have found many applications in crop measurement (Xia et al. 2015).

Chlorophyll is one of the key components in plant photosynthesis. Studies have found a high correlation between crop nitrogen levels and chlorophyll content when using spectral features in the range of 400–1000 nm (Ulissi et al. 2011; Lamb et al. 2002). Hence, many spectral sensors have been developed to measure plant chlorophyll content. Raper and Varco (2015) analyzed the multispectral reflectance and provided a simplified canopy chlorophyll content index to detect cotton chlorophyll, which indicated that reflectance in the red edge region strongly correlated with leaf nitrogen status. Thorp et al. (2015) estimated leaf chlorophyll with the partial least squares regression (PLSR) approach; the results showed the performance was better than NDVI and the physiological reflectance index. Inoue et al. (2016) compared the canopy chlorophyll contents of different plant types and regional scales and found the ratio spectral index with the reflectance at 815 nm and 704 nm was robust to predict canopy chlorophyll content. Recently, fluorescence sensors became another promising method for measuring plant chlorophyll and nitrogen content. Yang et al. (2015, 2016) used the ultraviolet (UV) laser to induce fluorescence and measured the intensity of fluorescence peaks at 685 nm and 740 nm to estimate paddy rice nitrogen content with back-propagation neural networks and support vector machine (SVM) models. They found that the intensity of fluorescence peaks was more sensitive and accurate than the fluorescence ratios in estimating nitrogen content. Longchamps and Khosla (2014) also conducted some tests for the multiplex fluorescence sensor and verified that fluorescence sensors can measure the variation of chlorophyll at the early stages of plant growth.

Crop water stress is an important parameter for assessing plant health. Stomatal conductance and leaf water potential (LWP) are vital indicators of plant water stress, and canopy temperature is a surrogate for stomatal conductance (Prashar and Jones 2016). Crop water stress index (CWSI) is successfully related to LWP (Bellvert et al. 2015). Water stress can be measured through canopy temperature or feature response of water in near-infrared or far-infrared reflectance spectra of canopy. Stationary temperature probes can be installed near canopy to log long-term canopy temperature data, e.g., SmartField (smartfield.com). With the decrease of their price, thermal cameras, e.g., FLIR camera (flir.com), have been used to evaluate water stress of crop in large-scale fields. They measure radiation from target objects in the 8–14 μm wavelengths and are often called longwave thermal infrared cameras. Ondimu and Murase (2008) developed a thermal-reflectance imaging system (TRIS) for detecting water stress in Sunagoke moss.

2.3 Soil Sensing Technology

Soil health is the continued capability of soil to function as a vital living ecosystem that sustains plants, animals, and humans. This is the result of the status and interaction of soil biological, chemical, and physical properties. Inspecting soil conditions has been routine operations for farmers to make decisions of planting, fertilizing, irrigation, and other agricultural operations. Traditionally, many soil physical and chemical properties are tested in dedicated laboratories, which may take a long time and high costs. Real-time, in-field soil property sensors have always been in demand for long-term, continuous soil monitoring. Most soil sensing technologies developed over the past years for field applications are optical, electrochemical, or dielectric in nature. Some only measure soil moisture, while others may be able to measure multiple parameters simultaneously.

Recently, precision irrigation techniques offer the potential to reduce agricultural water consumption without adversely affecting agricultural production. In precision irrigation, soil moisture is one of the key variables to calculate crop water demand. It is also one of the important environmental factors in controlling water and heat energy exchange between land surface and atmosphere through evaporation and plant transpiration (Jackson et al. 2007). The amount of water contained in a unit mass or volume of soil and the energy state of water in the soil are important factors affecting the growth of plants and yields. The demand from precision irrigation applications has motivated a significant increase in the types of commercially available devices for soil moisture content (SMC) measurement such as time-domain reflectometry (Huisman et al. 2001, 2002) and ground penetration radar (Lambot et al. 2006). A dielectric-based moisture sensor was evaluated under dynamic conditions by incorporating it into a nylon block that was attached to an instrumented tine (Liu et al. 1996). Results indicated the feasibility of this approach for real-time applications. This type of sensor has the advantages of low cost and very fast response. This moisture sensor not only responded to soil moisture but also salinity, soil texture, and temperature. Sensors with sensitivity to other parameters require signal compensation, adding significant cost and making them too expensive for use in a large-scale, networked measurement system. Capacitance probe sensors are a popular electromagnetic method for estimating soil water content. The basic principle is to incorporate the soil into an oscillator circuit and measure the resonant frequency. Capacitance probes are relatively cheap, safe, easy to operate, energy efficient, and easily automated (Kelleners et al. 2004). Many commercially available sensors are used to measure SMC at a single location in a field. Data is collected by a technician using a portable data acquisition device or by a stationary data logger with a preset sampling frequency. In order to sample SMC at multiple spots in a field, the technician needs to take the measurements one by one at each preselected spot. Alternatively, a number of stationary data acquisition devices need to be set up at the preselected locations and collect and store measured data to be downloaded during a technician's field visit. Wireless sensor network (WSN) is one of the newest and most promising technologies for soil moisture monitoring. The networked SMC measurement system allows in situ, timely measurements. It consists of multiple small sensor nodes, each of which has limited onboard signal processing ability and is equipped with one or more sensors and their signal conditioning circuits. The communication between sensor nodes is short-distanced and normally based on radio frequency (RF) (Wang et al. 2006; Wark et al. 2007). The in-field data are collected by a gateway or base station, which is equipped with a long-distance modem to relay the collect data to a remote data center.

Other soil sensors are available to measure multiple soil properties. A near-infrared (NIR) soil sensor measured soil spectral reflectance within the waveband of 1600–2600 nm to predict soil organic matter and moisture contents of surface and subsurface soils (Hummel et al. 2001). An online, real-time soil spectrophotometer measured soil spectral reflectance in the visible and NIR wavebands at a ground speed of 3.6 km/h. Field tests demonstrated linear relationships between reflectance at certain wavelengths and various soil properties, including soil organic matter and moisture content (Shibusawa et al. 2000). A soil electrical conductivity (EC) sensor based on a four-electrode method has proven effective in detecting several yield-limiting factors in nonsaline soils (Lund et al. 2000). In France, an eight-rolling electrode sensor was developed to measure soil EC at three depths (Dabas et al.

2000). Combining a soil EC probe with an automated penetrometer, soil subsurface can be mapped (Drummond et al. 2000). A soil EC sensor using the electromagneticinduction method is a noncontact sensor. EC measured using this sensor correlated well with a soil productivity index, which combines effects of bulk density, water-holding capacity, salt, and pH (Myers et al. 2000). Soil moisture content sensors were designed under different physical principles, including time-domain reflectivity (TDR), standing-wave ratio (Sun et al. 1999), and depolarization of a laser light (Zhang and Taylor 2000). A ground-penetrating radar was used to produce a contour map to indicate clay lenses, which govern the magnitude and direction of groundwater movement (Dulaney et al. 2000).

2.4 Root Sensing Technology

Root is an essential plant organ below ground and directly influences the absorption and usage of nutrient and water for plant growth. In general, plants respond to ambient biotic and abiotic stresses based on their root system architecture (RSA). Especially, under drought stress, an appropriate RSA can greatly promote the adaptation of plants against various extreme weather and climate changes. The main parameters of roots that are measured are total root lengths, primary root lengths, number of lateral roots, network width, root volume, root surface areas, etc. (Wasaya et al. 2018; Lee 2016). However, measuring RSAs is a challenging task due to their invisibility. Traditional methods for root measurement rely on digging out whole plant roots and cleaning incidental soil and manually measuring morphologic parameters. This process is often laborious, time-consuming, and costly (Trachsel et al. 2010).

Recently new methods for RSA evaluation are investigated and tested under controlled environment and field conditions. Under the controlled environment, plants were usually cultivated in transparent containers with clear media so that almost complete root structure can be visualized and measured with imaging techniques. Color digital cameras or scanners can be used to measure plant roots at various growth stages. This approach is widely implemented combined with soil-free techniques, such as hydroponics, aeroponics, gel plates, and growth pouches (Atkinson et al. 2019). Some researchers obtained a whole root structure using 3D reconstruction from several 2D side-view images that were obtained at various angles. Han and Yan (2018) developed a 3D imaging system using simple space carving to quantify the RSA of rice. Several software packages were reported to be able to extract root traits using 3D image reconstruction, e.g., RootReader 3D (Pineros et al. 2016) and DynamicRoots (Jiang et al. 2018; Symonova et al. 2015).

Ground-penetrating radar (GPR) is a commercially available method that uses radar pulses to image the subsurface. This nondestructive method uses electromagnetic radiation in high-frequency ranges (usually 10 MHz to 2.6 GHz) to detect the reflected signals from subsurface structures. They are used to measure tree root systems, which are usually larger than crop root systems.

Nondestructive root evaluation remains to be a very challenging task. Computer tomography with X-ray (CT scan) is a radiation-based technique, which enables the visualization of the interior of solid objects. It can be used to retrieve 3D plant root architecture based on the different attenuations of roots, air pores, water, soil, and other media. Maenhout et al. (2019) tested a method to segment large mature maize roots from X-ray CT images. The method is able to remove the soil particles that are around plant roots and extract root volumes. However, several factors can affect the measurements when using CT scan. First, the medium plays an important role in segmenting roots in CT images. X-ray attenuation values vary with soil type, soil moisture content, air-filled pores, root water status, root material, and age, which lead to the fact that roots cannot be extracted accurately just using threshold-based methods (Rogers et al. 2016). Second, some organic matters in soil will be mixed with plant roots, which leads to overlap among the attenuation values of the soil and roots. Similar to CT scan, magnetic resonance imaging (MRI) is another potential method to visualize 3D structure of plant roots under field conditions (Koch et al. 2019). Dusschoten et al. (2016) estimated the RSA of maize and quantified the growth of roots using MRI. Pflugfelder et al. (2017) tested and analyzed the effects of natural soil substrates and soil moisture on MRI root image qualities. The results showed that the two factors have no effect on image quality in the range of 50–80% water-holding capacity. However, up to now, there is still no mature method for nondestructive crop root measurement under field conditions.

2.5 Platforms for Sensing Technologies

Various types of platforms have been used to meet the requirements of different crop and soil sensing applications. In-field sensor networks, ground mobile platforms, manned and unmanned aerial vehicles, and satellites have all been used.

2.5.1 Ground-Based Platforms

Wireless sensors have been used in precision agriculture to assist in (1) spatial data collection, (2) precision irrigation, (3) variable-rate technology, and (4) supplying data to farmers. A mobile field data acquisition system was developed by Gomide et al. (2001) to collect data for crop management and spatial variability studies. The system consisted of a data collection vehicle, a manager vehicle, and data acquisition and control systems on farm machines. The system was able to conduct local field survey and to collect data of soil water availability, soil compaction, soil fertility, biomass yield, leaf area index, leaf temperature, leaf chlorophyll content, plant water status, local climate data, insect-disease-weed infestation, and grain yield. The data collection vehicle retrieved data from farm machines via a WLAN and analyzed, stored, and transmitted the data to the manager vehicle wirelessly. The operator in the vehicle monitored the performances of the farm machines and the data acquisition systems and troubleshot problems based on received data. Mahan and Wanjura (2004) cooperated with a private company to develop a wireless, infrared thermometer system for in-field data collection. The system consisted of infrared sensors, programmable logic controllers, and low-power radio transceivers to collect data in the field and transmit it to a remote receiver outside the field. Various environmental information can be collected by the in-field sensor networks, such as conditions of light, wind, air, soil, etc. For light information, solar radiation and light intensity are the main parameters. For wind conditions, wind speed and direction are the parameters of interest. Air humidity, temperature, atmospheric pressure, and rainfall are often collected and monitored. Soil moisture, temperature, dielectric permittivity, pH, conductivity, and salinity are collected to evaluate soil conditions. Meanwhile, plant information can also be collected by in-field sensor networks directly. Sensors for photosynthesis, carbon dioxide (CO_2) concentration, hydrogen, leaf wetness, leaf temperature, leaf area index (LAI), canopy conductance, evaporation rate, and stem moisture have been used as stationary nodes of in-field sensor networks. The data can be relayed to a remote computer to assist in making management decisions.

2.5.2 Ground Mobile Platform

Ground mobile platform is another category of ground-based platforms. Ground mobile platforms used Global Positioning System (GPS) to gain location information for every data collected. Compared with other platforms, ground mobile platform has two distinct features: high mobility and high flexibility. These features endow it with the best data collection power among all outdoor platforms. With high mobility, ground mobile platforms can scout a large area and collect crop and soil data in a relatively short time duration. Meanwhile, high flexibility lets ground mobile platforms move in-field smartly and carry various sensors to make close observations of crops. Close observation consists of both top-view and side-view observations for crops. Other than the commonly used top-view observation among all platforms, close side-view observation on ground mobile platforms can get under canopy sights.

Based on different power sources and control patterns, ground mobile platform has different variants, such as trolley cart-based platforms (Crain et al. 2016), tractor-based platforms (Baker III et al. 2016), and mobile robot-based platforms (Young et al. 2018; Bao et al. 2019). Trolley cart-based platforms are constructed on specially selected or designed trolley carts, whose widths are multiples of crop row spacing. Therefore, their wheels move on the terrain between two adjacent crop rows and carry different sensors to collect plant information. Trolley cart-based platforms are often driven manually. As a result, their payload, endurance time, and coverage area are much smaller than engine-driven platforms. Tractor-based platforms are developed by adding various sensors on engine-driven agricultural

machinery. With a powerful engine, tractor-based platforms can carry more sensor/
battery payloads and have longer endurance time and larger scouting coverage area.
Mobile robot-based platforms employ mobile robots as sensor carriers. They can be
fully autonomous or controlled remotely. There are different configurations for the
mobile robots. Some are small and can run between two adjacent crop rows. Others
may run over multiple crop rows. By cruising autonomously, mobile robot-based
platforms can work day and night to collect plant data. Even if their moving speed
is not higher than tractor-based or aerial-based platforms, their longer working time
can still guarantee a large coverage in a relatively short time duration (Qiu
et al. 2019).

2.5.3   Aerial Platforms

Compared with ground-based platforms, aerial-based platforms have much stronger
capabilities in monitoring speed and area coverage. As a result, many researchers
choose to collect plant data with aerial-based platforms (Atkinson et al. 2018).
There are different types of aerial-based platforms for plant phenotyping, including
unmanned aerial vehicles (Sankaran et al. 2015), manned aircraft (Gonzalez-Dugo
et al. 2015), and satellites (Lobell et al. 2015). Different types of aerial-based platforms have different features and advantages. Unmanned aerial vehicles can get
high-resolution data, manned aircrafts have high scouting speeds, satellites have
large monitoring coverage, and blimps can conduct low speed monitoring at higher
altitudes. But they are all sensitive to weather conditions, such as strong wind,
heavy cloud, and rainstorm. Furthermore, airspace regulations constrain the application area of most aerial-based platforms.

2.6   Examples of Sensing Technologies for Precision
Agriculture Applications

In this section, we report on two projects at Oklahoma State University, aimed at
advancing sensing systems for precision agriculture. The first one is focused on the
development of a wireless sensor network for in-field soil moisture monitoring. The
second one is focused on the development of sensing systems for in-field peanut
phenotyping.

2.6.1   A Wireless Sensor Network for In-Field Monitoring

Li et al. (2011) reported a hybrid soil sensor network (HSSN) designed and deployed
for in situ, real-time soil property monitoring (Fig. 2.1). The HSSN included a local
wireless sensor network (LWSN) formed by multiple sensor nodes installed at

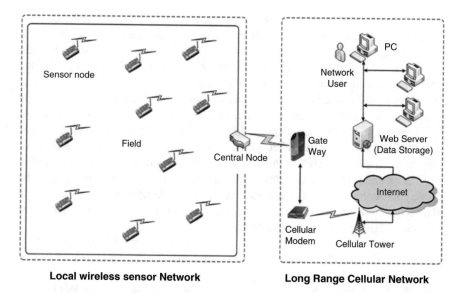

Local wireless sensor Network   **Long Range Cellular Network**

Fig. 2.1 A hybrid soil sensor network (HSSN) for in situ, real-time soil property monitoring (Li et al.2011)

preselected locations in the field to acquire readings from soil property sensors buried underground at four depths and transmit the data wirelessly to a data sink installed on the edge of the field and a long-distance cellular communication network (LCCN). The field data were transmitted to a remote web server through a GPRS data transfer service provided by a commercial cellular provider. The data sink functioned as a gateway, which received data from all sensor nodes, repacked the data, buffered the data according to the cellular communication schedule, and transmitted the data packets to LCCN. A web server was implemented on a PC to receive, store, process, and display the real-time field data. Data packets were transmitted based on an energy-aware self-organized routing algorithm.

The experiments were conducted at a wheat field of an experimental farm. The field was prepared and evenly divided into ten strips, each with a size of approximately 10 m by 30 m. The LWSN included a central node and ten sensor nodes working together to collect real-time soil property data. Since the LWSN was intended for unsupervised in-field applications, the wireless network had to be simple but robust enough to operate in harsh conditions without excessive human interference. A single-hop networking cluster with star topology allowed direct communication between the central node and each sensor node inside one cluster without complex routing and routing table maintenance, which usually requires a 5-min interval. This brings great flexibility in network scale extension since a sensor node could be added into one cluster without routing table reconstruction, or a new cluster could be deployed adjacently without interfering with the original ones. Another advantage is robustness; a failed sensor node won't influence the whole network. The LRCN consisted of a gateway, a cellular modem, and a remote web server. Except for the web server, all the WSN components were installed in the experimental field. Each sensor node was installed on the edge of the strip. The soil property sensors were distributed into the strips with physical locations predefined before deployment based on previous knowledge of soil variability and stayed stable unless a network reconfiguration was implemented. The central node and the gateway were installed at one end of the strip located in the center of the field. At each location, a sensor node acquired hourly soil property data including volume metric soil moisture content (VWC), electrical conductivity (EC), and near-surface temperature and wirelessly transmitted them to the central node for temporary storage. The central node uploaded stored data to a gateway based on a request from the gateway through wireless communications. Received packets on the gateway were then relayed to a cellular modem and transmitted to a web server through cellular network and GPRS service and the Internet.

In total, ten sensor nodes were deployed in an experimental field. Each sensor node included four soil property sensors buried at four depths of 5.08 cm, 15.24 cm, 30.48 cm, and 60.96 cm underground, a data acquisition unit, and a transceiver. Five out of the ten sensor nodes used only EC-5 soil moisture sensors (ECH2O probe, Decagon Devices, Pullman, MA, USA) to measure VWC at the four depths and five locations. Each of the other five sensor nodes had EC-5 sensors installed at the lower three depths and an EC-TE sensor (ECH2O-TE, Decagon Devices, Pullman, MA, USA) at the top. Hence, each of these sensor nodes could collect multiple soil properties including soil volumetric water content at four depths and temperature and bulk electrical conductivity (EC) at the depth of 5.08 cm. In order to implement real-time soil property monitoring, software for the sensor nodes, the gateway, and the web server were developed. Communication protocols were also developed with the considerations of large energy-saving, high success rate of packet delivery and low path-loss communications.

Field tests showed that the developed wireless sensor network was able to acquire real-time soil property data including volumetric water content, electrical conductivity, and temperature at different underground depths and various locations. A long-range cellular network and Internet services were able to relay the field data to a remote database on a web server for data storage and user querying. The data packet delivery rate was above 90% for most of the nodes.

2.6.2 *In-Field Peanut Phenotyping Systems*

Canopy architecture of plants affects solar radiation interception, plant growth, and ability to compete with weeds. Open canopies and upright growth habits reduce disease incidence by creating microclimates that are less conducive to pathogen growth or by reducing opportunities for plant contact with infested soil. Although canopy architecture has profound effects in plant growth, it is hard to define and quantify a three-dimensional structure and thus is rarely measured or adopted in most crop breeding and management programs.

Yuan et al. (2017) developed a ground-based phenotyping platform to evaluate peanut canopy architecture. The platform consisted of an onboard LiDAR sensor (LMS291-S05, SICK AG, Waldkirch, Germany), a video camera (GoPro 4, GoPro Inc., California, USA), a shaft encoder (Dynapar, Gurnee, Illinois, USA) attached to one of the rear wheels to record the location stamp of every LiDAR scan, a laptop, and a battery unit (24 V, 18.0 Amp. Hr.) (see Fig. 2.2). The LiDAR was mounted on the platform at 1 m above ground and was oriented to face downward. To obtain a reasonable field of view of the entire canopy within a single row, the LiDAR was configured to operate in a continuous line scan mode with a field of view of 100° (2.38 m in width) and a resolution of 0.25°. The laser scanner output had a total of 401 points for every line scan. To ensure high-speed data collection (500 kbps/s), the LiDAR was connected to a laptop through a serial-to-Ethernet converter (DeviceMaster 500, Comtrol Co., New Brighton, Minnesota, USA). A data acquisition program was developed using LabVIEW (the LabVIEW 2011, National Instruments Co., Austin, Texas, USA) to communicate with the LiDAR, correctly receive data packages, extract and convert distance data from polar to Cartesian coordinates, and save data as an MS Excel file. All line-scanned data were stored with a distance stamp and a time stamp. A video camera was mounted 10 cm from the center plane of the LiDAR and oriented downward to obtain a similar field of view (FOV). The camera recorded videos during field tests to verify LiDAR measurements in the field.

A software package was developed to handle data preprocessing, image processing, feature extraction, and statistical analysis. The data preprocessing included plant height calculation, identification of region of interest, boundary determination for a target plot, and noisy and outlier removal. The line scan data from the LiDAR

Fig. 2.2 Measurement system and experimental field setup: (**a**) Ground-based mobile data acquisition system; (**b**) four-row, 12-plot field setup: R3 was the measured row. The arrows show the travel direction of the measurement system (Yuan et al. 2019)

were combined into images and processed using image processing algorithms. The canopy height, width, and shape/density were then extracted and evaluated. Feature indices were defined and calculated from the images and used to describe the shape of the peanut canopies.

Field experiments were conducted at Oklahoma State University's Caddo Research Station in Fort Cobb, Oklahoma, USA. Three peanut cultivars of the runner market type were planted: Georgia-04S, McCloud and Southwest Runner . For easier planting, each plot had four rows but only one row was measured. This plot setup permitted the designed platform to scan one row without treading on the plants within the scanned row. Each plot was 3.66 m in width and 4.57 m in length. The experimental design was a randomized complete block design with four replications for each cultivar. The 12 four-row-wide plots were arranged in a single line in the field (Fig. 2.2b), and plots were separated by 1.52 m borders (Fig. 2.2a). To provide fixed reference points within each plot over the collection periods, 0.76 m-long metal posts were installed within the center length of each scanned row at 0.9 m, 1.8 m, 2.7 m, and 3.7 m.

The results showed that the developed LiDAR-based system was an effective tool for assessing peanut canopy architecture under field conditions. The developed algorithms were able to extract features of peanut canopy architecture, specifically canopy height, width, and shape/density. The descriptors used to quantify canopy shape and density, i.e., Euler number, entropy, cluster count, and mean area of connected objects, were effective for describing canopy characteristics and for discriminating among different cultivars. Finally, the canopy height data collected by LiDAR was highly correlated with ground-truth measurements with a R^2 of 0.915.

References

Atkinson, J.A., R.J. Jackson, A.R. Bentley, E. Ober, D.M. Wells, (2018). Field Phenotyping for the Future. Annual Plant Reviews, Issue 3.

Atkinson, J. A., Pound, M. P., Bennett, M. J., & Wells, D. M. (2019). Uncovering the hidden half of plants using new advances in root phenotyping. *Current Opinion in Biotechnology, 55*, 1–8.

Baker, J., III, Zhang, N., Sharon, J., Steeves, R., Wang, X., Wei, Y., & Poland, J. (2016). Development of a field-based high-throughput mobile phenotyping platform. *Computers and Electronics in Agriculture, 122*, 74–85.

Bao, Y., Tang, L., Breitzman, M. W., Fernandez, M. G. S., Schnable, P. S. (2019). Field-based robotic phenotyping of sorghum plant architecture using stereo vision. *Journal of Field Robotics, 36*, 397–415.

Bellvert, J., Marsal, J., Girona, J., & Zarco-Tejada, P. J. (2015). Seasonal evolution of crop water stress index in grapevine varieties determined with high-resolution remote sensing thermal imagery. *Irrigation Science, 33*(2), 81–93.

Bulanon, D. M., Kataoka, T., Okamoto, H., & Hata, S. (2004). *Determining the 3-D location of the apple fruit during harvest. Automation technology for off-road equipment* (ASAE Number: 701P1004). St. Joseph: The American Society of Agriculture Engineers.

Crain, J. L., Wei, Y., Barker, J., III, Thompson, S. M., Alderman, P. D., Reynolds, M., Zhang, N., & Poland, J. (2016). Development and deployment of a portable field phenotyping platform. *Crop Science, 56*(3), 965–975.

Dabas, M., Boisgontier, D., Tabbagh, J., & Brisard, A. (2000, July 16–19). Use of a new sub-metric multi-depth soil imaging system (MuCEp c). In *Proceedings of fifth international conference on precision agriculture (CD)*. Bloomington: American Society of Agronomy/Crop Science Society of America/Soil Science Society of America.

Drummond, P. E., Christy, C. D., & Lund, E. D. (2000, July 16–19). Using an automated penetrometer and soil EC probe to characterize the rooting zone. In *Proceedings of fifth international conference on precision agriculture (CD)*. Bloomington: American Society of Agronomy/Crop Science Society of America/Soil Science Society of America.

Dulaney, W. P., Daughtry, C. S. T., Walthall, C. L., Gish, T. J., Timlin, D. J., & Kung, K. J. S. (2000, July 16–19). Use of ground-penetrating radar and remotely sensed data to understand yield variability under drought conditions. In *Proceedings of fifth international conference on precision agriculture (CD)*. Bloomington: American Society of Agronomy/Crop Science Society of America/Soil Science Society of America.

Dusschoten, D. V., Metzner, R., Kochs, J., Postma, J. A., Pflugfelder, D., & Bühler, J. (2016). Quantitative 3D analysis of plant roots growing in soil using magnetic resonance imaging. *Plant Physiology, 170*(3), 1176–1188.

Gibbons, G. (2000). *Turning a farm art into science: An overview of precision farming.* http://www.precisionfarming.com

Gomide, R. L., Inamasu, R. Y., Queiroz, D. M., Mantovani, E. C., & Santos, W. F. (2001). *An automatic data acquisition and control mobile laboratory network for crop production systems data management and spatial variability studies in the Brazilian center-west region* (ASAE Paper No.: 01-1046). St. Joseph: The American Society of Agriculture Engineers.

Gonzalez-Dugo, V., Hernandez, P., Solis, I., & Zarco-Tejada, P. J. (2015). Using high-resolution hyperspectral and thermal airborne imagery to assess physiological condition in the context of wheat phenotyping. *Remote Sensing, 2015*(7), 13586–13605.

Han, T., & Yan, F. K. (2018). Developing a system for three-dimensional quantification of root traits of rice seedlings. *Computers and Electronics in Agriculture, 152*, 90–100.

Huisman, J. A., Sperl, C., Bouten, W., & Verstraten, J. M. (2001). Soil water content measurements at different scales: Accuracy of time domain reflectometry and ground-penetrating radar. *Journal of Hydrology, 245*(1), 48–58.

Huisman, J. A., Snepvangers, J. J. J. C., Bouten, W., & Heuvelink, G. B. M. (2002). Mapping spatial variation in surface soil water content: Comparison of ground-penetrating radar and time domain reflectometry. *Journal of Hydrology, 269*(3), 194–207.

Hummel, J. W., Sudduth, K. A., & Hollinger, S. E. (2001). Soil moisture and organic matter prediction of surface and subsurface soils using a NIR sensor. *Computers and Electronics in Agriculture, 32*, 149–165.

Inoue, Y., Guérif, M., Baret, F., Skidmore, A., Gitelson, A., & Schlerf, M. (2016). Simple and robust methods for remote sensing of canopy chlorophyll content: A comparative analysis of hyperspectral data for different types of vegetation. *Plant, Cell and Environment, 39*(12), 2609–2623.

Jackson, T., Mansfield, K., Saafi, M., Colman, T., & Romine, P. (2007). Measuring soil temperature and moisture using wireless MEMS sensors. *Journal of Measurement, 41*(4), 381–390.

Jiang, N., Floro, E., Bray, A. L., Laws, B., Duncan, K. E., & Topp, C. N. (2018). High-resolution 4D spatiotemporal analysis of maize roots. *The Plant Cell.* https://doi.org/10.1101/381046.

Jones, C. L., Maness, N. O., Stone, M. L., & Jayasekara, R. (2004). *Sonar and digital imagery for estimating crop biomass* (ASABE Paper No. 043061). St. Joseph: The American Society of Agriculture Engineers.

Kelleners, T. J., Soppe, R. W. O., Ayars, J. E., & Skagg, T. H. (2004). Calibration of capacitance probe sensors in a saline silty clay soil. *Soil Science Society of America Journal, 68*, 770–778.

Koch, A., Meunier, F., Vanderborght, J., Garré, S., Pohlmeier, A., & Javaux, M. (2019). Functional–structural root-system model validation using a soil MRI experiment. *Journal of Experimental Boltany, 70*(10), 2797–2809.

Lamb, D. W., Steyn-Ross, M., Schaare, P., Hanna, M. M., Silvester, W., & Steyn-Ross, A. (2002). Estimating leaf nitrogen concentration in ryegrass (Lolium spp.) pasture using the chlorophyll red-edge: Theoretical modelling and experimental observations. *International Journal of Remote Sensing, 23*(18), 3619–3648.

Lambot, S., Weihermüller, L., Huisman, J. A., Vereecken, H., Vanclooster, M., & Slob, E. C. (2006). Analysis of air-launched ground-penetrating radar techniques to measure the soil surface water content. *Water Resources Research, 42*, W11403.

Lee, N. (2016). *High-throughput phenotyping of above and below ground elements of plants using feature detection, extraction and image analysis techniques*. MSc thesis, Iowa State University.

Li, Z., Wang, N., Taher, P., Godsey, C., Zhang, H., & Li, X. (2011). Practical deployment of an in-field soil property wireless sensor network. *Computer Standards & Interfaces, 36*(2), 278–287.

Liu, W., Upadahyaya, S. K., Kataoka, T., & Shibusawa, S. (1996). Development of a texture/soil compaction sensor. In *Proceedings of the 3rd international conference on precision agriculture* (pp. 617–630). Minneapolis: American Society of Agronomy.

Lobell, D. B., D. Thau, C. Seifert, E. Engle, B. Little, (2015). A scalable satellite-based crop yield mapper. Remote Sensing of Environment, 164: 324–333.

Longchamps, L., & Khosla, R. (2014). Early detection of nitrogen variability in maize using fluorescence. *Journal of Agronomy, 106*(2), 511.

Lund, E. D., Christy, C. D., & Drummond, P. E. (2000, July 16–19). Using yield and soil electrical conductivity (EC) maps to derive crop production performance information. In *Proceedings of fifth international conference on precision agriculture (CD)*. Bloomington: American Society of Agronomy/Crop Science Society of America/Soil Science Society of America.

Maenhout, P., Sleutel, S., Xu, H., Hoorebeke, L. V., Cnudde, V., & Neve, S. D. (2019). Semi-automated segmentation and visualization of complex undisturbed root systems with X-ray μCT. *Soil and Tillage Research, 192*, 59–65.

Mahan, J., & Wanjura, D. (2004). Upchurch, design and construction of a wireless infrared thermometry system. *The USDA annual report*. Project Number: 6208-21000-012-03. May 01, 2001–September 30, 2004.

Myers, D. B., Kitchen, N. R., Miles, R. J., & Sudduth, K. A. (2000, July 16–19). Estimation of a soil productivity index on claypan soils using soil electrical conductivity. In *Proceedings of fifth international conference on precision agriculture (CD)*. Bloomington: American Society of Agronomy/Crop Science Society of America/Soil Science Society of America.

Ondimu, S., & Murase, H. (2008). Water stress detection in Sunagoke moss (Rhacomitrium canescens) using combined thermal infrared and visible light imaging techniques. *Biosystems Engineering, 100*(1), 4–13.

Pflugfelder, D., Metzner, R., Dusschoten, D. V., Reichel, R., Jahnke, S., & Koller, R. (2017). Non-invasive imaging of plant roots in different soils using magnetic resonance imaging (MRI). *Plant Methods, 13*, 102.

Pineros, M. A., Larson, B. G., Shaff, J. E., Schneider, D. J., Falcão, A. X., & Yuan, L. (2016). Evolving technologies for growing, imaging and analyzing 3D root system architecture of crop plants. *Journal of Integrated Plant Biology, 58*(3), 230–241.

Prashar, A., & Jones, H. G. (2016). Assessing drought responses using thermal infrared imaging. *Methods in Molecular Biology, 1398*, 209–219.

Qiu, Q., Sun, N., Bai, H., Wang, N., Fan, Z. Q., Wang, Y. J., Meng, Z. J., Li, B., & Cong, Y. (2019). Field-based high-throughput phenotyping for maize plant using 3D LiDAR point cloud generated with a "phenomobile". *Frontiers in Plant Science, 10*, 554.

Raper, T. B., & Varco, J. J. (2015). Canopy-scale wavelength and vegetative index sensitivities to cotton growth parameters and nitrogen status. *Journal of Precision Agriculture, 16*(1), 62–76.

Rogers, E. D., Monaenkova, D., Mijar, M., Nori, A., Goldman, D. I., & Benfey, P. N. (2016). X-ray computed tomography reveals the response of root system architecture to soil texture. *Plant Physiology, 171*, 2028–2040.

Rovira-Mas, F., Zhang, Q., & Reid, J. F. (2003). *Stereo 3D crop maps from aerial images* (ASABE Paper No. 031003). St. Joseph: The American Society of Agriculture Engineers.

Saeys, W., Lenaerts, B., Craessaerts, G., & Baerdemaeker, J. D. (2009). Estimation of the crop density of small grains using Lidar sensors. *Biosystems Engineering, 102*, 22–30.

Sankaran, S., Khot, L. R., Espinoza, C. Z., Jarolmasjed, S., Sathuvalli, V. R., Vandemark, G. J., Miklas, P. N., Carter, A. H., Pumphrey, M. O., Knowles, N. R., & Pavek, M. J. (2015). Low-altitude, high-resolution aerial imaging systems for row and field crop phenotyping: A review. *European Journal of Agronomy, 70*, 112–123.

Sharma, L. K., Bu, H., Franzen, D. W., & Denton, A. (2016). Use of corn height measured with an acoustic sensor improves yield estimation with ground based active optical sensors. *Computers and Electronics in Agriculture, 124*, 254–262.

Shi, Y., Wang, N., Taylor, R. K., Raun, W. R., & Hardin, J. A. (2013). Automatic corn plant location and spacing measurement using laser line-scan technique. *Journal of Precision Agriculture, 4*(5), 478–494.

Shi, Y., Wang, N., Taylor, R. K., & Raun, W. R. (2015). Improvement of a ground-LiDAR-based corn plant population and spacing measurement system. *Computers and Electronics in Agriculture, 112*, 92–101.

Shibusawa, S. (1998, October 20–22). Precision farming and terra-mechanics. In *The 5th ISTVS Asia-Pacific regional conference.* Korea.

Shibusawa, S., Anom, W. S., Sato, H., & Sasao, A. (2000, July 16–19). On-line real-time soil spectrophotometer. In *Proceedings of fifth international conference on precision agriculture (CD)*. Bloomington: American Society of Agronomy/Crop Science Society of America/Soil Science Society of America.

Sui, R., Thomasson, J., & Ge, Y. (2012). Development of sensor systems for precision agriculture in cotton. *International Journal of Agricultural and Biological Engineering, 4*(5), 1–14.

Sui, R., Fisher, D. K., & Reddy, K. N. (2013). Cotton yield assessment using plant height mapping system. *The Journal of Agricultural Science, 5*(1), 23–31.

Sun, Y., Wang, M., & Zhang, N. (1999). *Measuring soil water content using the principle of standing-wave ratio* (ASAE Paper No. 00-3127). St. Joseph: American Society of Agricultural Engineers.

Swain, K. C., Zaman, Q. U., Schumann, A. W., & Percival, D. C. (2009). *Detecting weed and bare-spot in wild blueberry using ultrasonic sensor technology* (ASABE Paper No. 096879). St. Joseph: American Society of Agricultural Engineers.

Symonova, O., Topp, C. N., & Edelsbrunner, H. (2015). DynamicRoots: A software platform for the reconstruction and analysis of growing plant roots. *PLoS One, 10*(6), e0127657. https://doi.org/10.1371/journal.pone.0127657.

Thorp, K. R., Gore, M. A., Andrade-Sanchez, P., Carmo-Silva, A. E., Welch, S. M., White, J. W., & French, A. N. (2015). Proximal hyperspectral sensing and data analysis approaches for field-based plant phenomics. *Computers and Electronics in Agriculture, 118*, 225–236.

Trachsel, S., Kaeppler, S. M., Brown, K. M., & Lynch, J. P. (2010). Shovelomics: High throughput phenotyping of maize (Zea mays L.) root architecture in the field. *Plant and Soil, 341*, 75–87.

Ulissi, V., Antonucci, F., Benincasa, P., Farneselli, M., Tosti, G., & Guiducci, M. (2011). Nitrogen concentration estimation in tomato leaves by VIS-NIR non-destructive spectroscopy. *Sensors, 11*(12), 6411–6424.

Wang, N., Zhang, N., & Wang, M. (2006). Wireless sensors in agriculture and food industry: Recent developments and future perspective. *Computers and Electronics in Agriculture, 50*(1), 1–14.

Wark, T., Corke, P., Sikka, P., Klingbeil, L., Guo, Y., Crossman, P., Valencia, P., Swain, D., & Bishop-Herley, G. (2007). Transforming agriculture through pervasive wireless sensor networks. *Pervasive Computing, 6*(2), 50–57.

Wasaya, A., Zhang, X., Fang, Q., & Yan, Z. (2018). Root phenotyping for drought tolerance: A review. *Journal of Agronomy, 8*(11), 241.

Wei, J., & Salyani, M. (2004). Development of a laser scanner for measuring tree canopy characteristics: Phase 1. Prototype development. *Transactions of ASAE, 47*(6), 2101–2107.

Wei, J., & Salyani, M. (2005). Development of a laser scanner for measuring tree canopy characteristics: Phase 2. Foliage density measurement. *Transactions of ASAE, 48*(4), 1595–1601.

Xia, C., Wang, L., Chung, B., & Lee, J. (2015). In situ 3D segmentation of individual plant leaves using a RGB-D camera for agricultural automation. *Sensors, 15*(8), 20463–20479.

Yang, J., Shi, S., Gong, W., Du, L., Ma, Y. Y., & Zhu, B. (2015). Application of fluorescence spectrum to precisely inverse paddy rice nitrogen content. *Plant, Soil and Environment, 61*(4), 182–188.

Young, S. N., E. Kayacan, J.M. Peschel, (2018). Design and field evaluation of a ground robot for high-throughput phenotyping of energy sorghum. Journal of Precision Agriculture, 20(4): 697–722.

Yang, J., Gong, W., Shi, S., Du, L., Sun, J., & Song, S. (2016). Analysing the performance of fluorescence parameters in the monitoring of leaf nitrogen content of paddy rice. *Scientific Reports, 6*, 28787.

Yuan, L., Z.Y. Bao, H.B. Zhang, Y.T. Zhang, X. Liang, (2017). Habitat monitoring to evaluate crop disease and pest distributions based on multi-source satellite remote sensing imagery. Optik, 145: 66–73.

Yuan, H., Bennett, R. S., Wang, N., & Chamberlin, K. D. (2019). Development of a peanut canopy measurement system using a ground-based LiDAR sensor. *Frontiers in Plant Science, 10*, 203.

Zhang, N., & Taylor, R. (2000, July 16–19). Applications of a field-level geographic information system (FIS) in precision agriculture. In *Proceedings of fifth international conference on precision agriculture (CD)*. Bloomington: American Society of Agronomy/Crop Science Society of America/Soil Science Society of America.

Dr. Man Zhang is a professor and Dean of the College of Information and Electrical Engineering, China Agricultural University; member of Key Laboratory of Modern Precision Agriculture System Integration Research, Ministry of Education, China; and member of the Asian Association of Precision Agriculture and the American Society of Agricultural and Biological Engineers (ASABE). In recent years, her research areas mainly focused on the automatic acquisition and processing of agricultural information and intelligent control and automatic navigation of agricultural machines. She has successively presided over three projects of the National Natural Science Foundation of China, one project of the Beijing Natural Science Foundation, and one project of the National Key R&D Programmes of China. She has also won the second prize of the National Science and Technology Progress Awards in 2008 and published more than 60 academic papers.

Dr. Ning Wang Research and teaching in sensors, controls, mechatronics, sensor networks, and data analytics for agricultural and food production and processes. (a) **Professional Preparation** China Agricultural University, Applied Electronic Technology, B.S., 1987, M.S. 1990 Asian Institute of Technology, Industrial Engineering, M.S., 1995. Kansas State University, Biological and Agricultural Engineering, Ph.D., 2002; Electrical Engineering, M.Sc., 2002. (b) **Appointments** Oklahoma State University: Associate Professor, 2010–present Oklahoma State University: Assistant Professor, 2006–2010 McGill University, Canada: Assistant Professor, 2002–2006. Kansas State University and USDA Grain Marketing and Production Research Center (GMPRC): Postdoctoral Researcher, 02-08/2002; Graduate Research Assistant, 1998–2002. Master Call Co. Ltd., Thailand: Technical Assistant to CEO, 1996–1997 China Agricultural University: Lecturer, 1990–1994.

(c) **Synergistic Activities (Selected)** (i) <u>Leadership at National Technical Society:</u> Currently serving as the Division Chair of the Information and Electrical Technology Division (2013–2014) of the American Society of Agricultural and Biological Engineers (ASABE). Served as the Chair of the Technical Committees of the IET Division, including Instrumentation and Control Committee (IET-353); Machine Vision Committee (IET-312); and Mechatronics and Biorobotics (IET-318). Initiator of the ASABE Robot Competition. (ii) <u>Editorship for Professional Journals:</u> Associate Editor for *Transactions of the ASABE* and *Applied Engineering in Agriculture,* 2003–present; Editorial Board Member, *International Journal on Bionic Engineering*, 2003–present; Associate Editor, *International Journal of Agricultural and Biological Engineering*, 2008–present. (iii) <u>Reviewer for Technical Journals and Grant Agencies</u>: Invited to review manuscripts for ten refereed journals and to review grant proposals for four agencies (USDA, NSERC Canada, CFI Canada, and Florida Citrus Advanced Technology Program (FCATP)). (d) **Most Recent Peer-Reviewed Publication** 1. Qiu, Q., N. Sun, H. Bai, N. Wang, Z. Fan, Y. Wang, Z., B. Li, B., and Y. Cong, 2019. Field-based high-throughput phenotyping for maize plant using 3D LIDAR point cloud generated with a "phenomobile". *Frontiers in Plant Science*. 10(2019): 554. 2. Qiu, G., E. Lü, N. Wang, H. Lu, F. Wang, F. Zeng, 2019. Cultivar Classification of Single Sweet Corn Seed Using Fourier Transform Near-Infrared Spectroscopy Combined with Discriminant Analysis. *Applied Sciences*, 9(8):1530. 3. Yuan, H., R.S. Bennett, N. Wang, K. D. Chamberlin, 2019. Development of a Peanut Canopy Measurement System Using a Ground-Based LiDAR Sensor, *Frontiers in Plant Science*. 10:203. 4. Zhai, C., J. Long, R. Taylor, P. Weckler, and N. Wang, 2019. Field Scale Row Unit Vibration Affecting Planting Quality. *Journal of Precision Agriculture*. https://doi.org/10.1007/s11119-019-09684-4. Published Sept 26, 2019. 5. Zhang, J., N. Wang, L. Yuan, F. Chen, and K. Wu, 2017. Discrimination of winter wheat disease and insect stresses using continuous wavelet features extracted from foliar spectral measurements. *Biosystems Engineering*. 162(2017): 20–29.

Dr. Liping Chen Institution: Beijing Research Center of Intelligent Equipment for Agriculture, Beijing Academy of Agriculture and Forestry Sciences (BAAFS), China. **Education:** B.Sc. (Agronomy), Beijing Agriculture University (now China Agriculture University), Beijing, P.R. China, 1991–1995. M.Sc. (Agronomy), China Agriculture University, Beijing, P.R. China, 1995–1998. Ph.D. (Agronomy), China Agriculture University, Beijing, P.R. China, 2000–2003. **Employment Record:** 1998–2000: Research Intern, Beijing Research Center for Information Technology in Agriculture, BAAFS. 2000–2005: Assistant Prof., Beijing Research Center for Information Technology in Agriculture, BAAFS. 2005–2008: Associate Prof., Beijing Research Center of Intelligent Equipment for Agriculture, BAAFS. 2008-2010: Professor, Beijing Research Center of Intelligent Equipment for Agriculture, BAAFS. 2010–2016: Professor, Deputy Director of Beijing Research Center of Intelligent Equipment for Agriculture, BAAFS. 2016-now: Professor, Director of Beijing Research Center of Intelligent Equipment for Agriculture, BAAFS. Dr. Liping Chen is the Director of Beijing Research Center of Intelligent Equipment for Agriculture, China National Research Center of Intelligent Equipment for Agriculture, and China National Engineering Laboratory of Agricultural Internet of Things. Dr. Liping Chen has been mainly engaged in research on precision agriculture information acquisition technology and equipment. She is Vice-President of the Beijing Society of Agricultural Informatization. She has served as Deputy Leader of the Application Working Group of the National Wireless Sensor Network Standard Committee, Expert of the National Agricultural Internet of Things Standard Working Group. Dr. Chen has obtained 55 national invention patents, published 92 papers, and won the three Second Class Prizes of the National Scientific and Technology Progress Award.

Chapter 3
Perspectives to Increase the Precision of Soil Fertility Management on Farms

Joann K. Whalen

Contents

3.1 Introduction: Why Did I Become a Soil Fertility Researcher?

My grandparents and parents farmed, so I always had a very strong connection to the land. From a young age, we helped to grow our own food and take care of animals. We went to exhibitions and competed for prizes for the best vegetables and were members of the local 4-H club. It seemed natural to go to an agricultural college, where I got my B.Sc. (Agr.) in agricultural chemistry.

In the final year of my BSc (Agr.) program, we had to do an Honors research project. My first attempted experiment aimed to evaluate potato growth with nitrogen fertilizer in a cultivated field versus in the nearby forest. The potatoes in the cultivated field gave better yields with higher rates of nitrogen fertilizer, which I expected, but those in the forest were the size of peas, regardless of how much nitrogen fertilizer I added. My family thought this was very funny, since it is obvious that we do not grow potatoes in the forest. Later, I understood that my result was due to the high nitrogen demand of the microorganisms and trees surrounding my forest potato plot – those potatoes never had a chance to get any nitrogen fertilizer because

J. K. Whalen (✉)
Department of Natural Resource Sciences, Macdonald Campus of McGill University,
Sainte-Anne-de-Bellevue, QC, Canada
e-mail: joann.whalen@mcgill.ca

© Springer Nature Switzerland AG 2021
T. K. Hamrita (ed.), *Women in Precision Agriculture*, Women in Engineering
and Science, https://doi.org/10.1007/978-3-030-49244-1_3

they could not compete with the forest biota! My experimental design was flawed because I neglected to replicate my experimental treatments, and we cannot understand very much from single observations (n = 1). Subsequently, I worked on a different Honors research project, which tested several types of food-waste compost as a soil amendment for tomato, involving analysis of the compost, soil, and tomato plants. This experiment was replicated (n = 3), so I could calculate the mean and standard error of each treatment and reach some conclusions by the end.

My best undergraduate course was in soil chemistry with Dr. Phil Warman. I liked it so much that I asked him for a chance to do a Master's thesis under his supervision, which he graciously accepted. I spent the next 18 months working intensively on soil sulfur chemistry and becoming very interested in arylsulfatase enzymes (Whalen and Warman 1996). After that, I joined the Soil Ecology group at Ohio State University for my PhD degree. This was a great opportunity because I got to work with Dr. Rob Parmelee, an exceptional mentor, and Dr. Clive Edwards, the world authority on earthworm ecology. During this time, I learned much about earthworm biology, ecology, and their contribution to nitrogen cycling using ^{15}N as a stable isotope tracer.

My PhD supervisors gave me a lot of independence to design experiments and make additional measurements to improve my project. In one experiment, I had to figure out a method to quantify earthworm feeding rates and track the amount of ^{15}N-labeled food that was assimilated into their body tissues. It turns out that earthworms are very messy eaters who will burrow into their food supply, given the chance, so I had to design a food-delivery system that they could not enter (Whalen and Parmelee 1999). My favorite experiment involved feeding earthworms with ^{15}N-labeled food, then freezing them, and putting their dead bodies into soil planted with ryegrass. We showed that a large proportion of ^{15}N moved from the earthworm into microbial biomass within 2 days and then was released from the microbial biomass and absorbed by the ryegrass within 4 days (Whalen et al. 1999). It was remarkable how quickly earthworm nitrogen became plant nitrogen. Follow-up studies on nitrogen cycling processes during my postdoctoral training with Dr. David Myrold and Dr. Peter Bottomley at Oregon State University taught me the value of density separation procedures for fractionating soil organic matter. We found that the undecomposed (light) fraction of soil organic matter has a high C/N ratio and immobilizes nitrogen, whereas the more, decomposed (heavy) fraction of soil organic matter is a net source of mineral nitrogen (Whalen et al. 2000).

My academic training during my MSc and PhD studies and as a postdoctoral fellow prepared me, in the best possible way, to be a better scientific communicator and a critical thinker. I am grateful that I had such a broad exposure to soil fertility, chemistry, biochemistry, and ecology during my education, because it has helped immensely to understand complex interactions in the soil-plant system.

3.2   Why Does Soil Fertility Matter?

Why should we be interested in soil fertility? The simple answer is that our global food supply depends upon the soil. In the United States alone, approximately 125 million hectares were planted with major field crops like corn, soybean, wheat, oats, barley, and other commodities during the 2019 growing season (Hubbs 2019). Worldwide production of staple grain crops including corn, wheat, rice, barley, sorghum, oats, and rye was about 2564 million metric tonnes in 2018–2019 (Statista 2020). Livestock like cattle, sheep, and goats graze across rangelands, which are extensive pastures of natural and improved vegetation on soil, extending across 3.6 billion hectares of land worldwide (Reid et al. 2014). By 2050, the world's population will be an estimated 9.7 billion individuals (UN DESA 2019), and all of those people will require nutritious food, every day of their lives.

Soil fertility researchers want to make sure the soil provides the right amount of essential macro- and micronutrients to the crop during its growth. Crops that are deficient in macronutrients – nitrogen, phosphorus, potassium, sulfur, calcium, and magnesium – are often stunted and do not reach their yield potential; in severe cases, a lack of macronutrients can result in crop failure. Micronutrients are needed in relatively small quantities, but plants must have adequate iron, manganese, boron, zinc, copper, chloride, molybdenum, and nickel to perform key functions such as photosynthesis, energy transfer reactions, and protein biosynthesis.

There is an easy solution to the problem – add fertilizer! However, solid fertilizer itself is rather useless to plants. Solid fertilizers must dissolve in water to release their nutrients into the soil solution, the water present in the pore space where plant roots grow. Furthermore, the nutrients need to be transformed into ions that move through the water-filled pore space to plant roots and are then absorbed into the roots. Plants do not discriminate among ions, which means that many fertilizers could potentially improve the soil fertility. If the solid fertilizer is granular triple superphosphate (calcium phosphate) purchased from a fertilizer company, then the water dissolves the fertilizer granule to release phosphate ($H_2PO_4^-$) ions that could be acquired by the root. Another solid fertilizer is chicken manure, which contains about 0.5% phosphorus (equivalent to 1.1% P_2O_5; Jones 2003), but mostly as organic phosphorus compounds like phytate. Because chickens do not produce the phytase enzyme in their digestive system, only 10%–30% of the organic phosphorus in their food is absorbed by the animal and most of it is excreted (Sims and Vadas 1997). When chicken manure is applied to soil, these organic phosphorus compounds are degraded by phosphatase enzymes produced by soil microorganisms and plant roots. In the presence of water, the phosphatase enzyme will hydrolyze part of the organic molecule to release phosphate (HPO_4^{2-}) ions into the soil solution (Fig. 3.1). If there is abundant water, the dissolution of granular triple superphosphate will occur faster than the hydrolysis of organic phosphorus compounds in chicken manure, which is why commercial fertilizers contain "fast-release" nutrients and organic fertilizers are considered a "slow-release" nutrient source.

Fig. 3.1 Water is needed to dissolve the inorganic phosphorus (P) in a commercial fertilizer like triple superphosphate. Water and phosphatase enzymes, produced by soil microorganisms and plant roots, are needed to hydrolyze organic P contained in organic fertilizers like chicken manure. Soluble $H_2PO_4^-$ ions are absorbed by fine plant roots and used for plant metabolic processes. Since it is easier to solubilize the inorganic P compounds than the organic P compounds, commercial products are generally considered to be "fast-release" fertilizer, and organic fertilizer is a relatively "slow-release" fertilizer. The numbers associated with each fertilizer are the guaranteed analysis, meaning the expected %N-%P_2O_5-%K_2O in the fertilizer. Triple superphosphate is a simple fertilizer that contains no nitrogen, 46% P_2O_5, and no potassium. Chicken manure is a compound fertilizer that contains 1.6% N, 1.1% P_2O_5, and 1% K_2O on average (Jones 2003)

The example in Fig. 3.1 illustrates the major dilemma in soil fertility research, which is how to manage soluble nutrients for the benefit of crops. When soil has a limited amount of soluble nutrients, plants cannot absorb enough ions to meet their requirements, and this puts the crop at risk of a nutrient deficiency. When soil contains an excess of soluble nutrients, we will have more than enough nutrients for the crop, and the rest of the soluble ions are susceptible to leach out of the soil or be transported through surface runoff into downstream aquatic systems. Both of these situations are highly undesirable.

In the next sections, I share some observations about soluble nutrients from my work as a soil fertility researcher. I will explain how these concepts are helping us to improve soil fertility management on farms, for greater precision in agriculture.

3.3   Nutrient-Deficient Soils

All soils, including those cultivated for agriculture, are inherently uneven and heterogeneous in their characteristics. Soil granolometry can change from coarse-textured (sandy) to fine-textured (clayey) across fields. Variation in the soil organic matter level may depend upon the amount of vegetation growing in a particular field and the decomposition rate of crop residues and other organic materials. Concentrations of plant-available nutrients may be abundant, sufficient, or insufficient for plant growth. These conditions are the result of the soil-forming processes that occur across landscapes and landforms, as well as agronomic practices that can alter the soil fertility on farms. In addition, the climatic conditions in a particular region will influence the ability of crops to acquire soluble nutrients.

Soil-forming factors, also known as pedogenic factors, are responsible for the inherent variability in soil properties because geologic processes are responsible for the type of parent material (i.e., bedrock) and the topographic features that occur in a particular field. The climate and the organisms (i.e., vegetation, biota) that have historically inhabited this area contribute to the physical and chemical weathering of the parent material and the overlying soil (Jenny 1980).

There are a few ways that the pedogenic processes may explain the presence of nutrient-deficient soils. First, the parent material may be impoverished in a particular mineral nutrient, so that the levels of that nutrient are inherently low. Iron chlorosis, resulting from inadequate iron nutrition in cereal crops, is common in soils that overlie calcareous parent material and consequently have low natural reserves of iron. In these high pH soils, the solubility of Fe^{2+} and Fe^{3+} is low because it forms precipitates with bicarbonate ions (CO_3^{2-}). Second, the weathering process may be intense due to abundant rainfall and humid conditions, such as in tropical oxisols. Abundant iron (III) and aluminum oxides and hydroxides in these soils have a high affinity for binding phosphate ions, resulting in widespread phosphate deficiency in such soils. Third, areas with sparse natural vegetation are likely to have a low organic matter content, which is the most important source of nitrogen and sulfur for plant nutrition and also contains organic phosphorus. Mineralization of soil organic matter is a natural process mediated by soil prokaryotes (bacteria and archaea) as well as soil fungi, which are responsible for biochemically transforming the organic nitrogen into the soluble ammonium (NH_4^+) and nitrate (NO_3^-) ions, the organic sulfur into soluble sulfate (SO_4^{2-}) ion, and the organic phosphorus compounds into soluble HPO_4^{2-}.

We can diagnose the reason for these and other nutrient deficiencies through soil testing and plant analysis. This tells us the reason for the deficiency. Then, we can select an appropriate fertilizer and apply it to correct the nutrient imbalance.

Another option is to select agronomic practices that increase the concentration of soluble nutrients. Modifying the soil pH with acidifying substances, to reduce the soil pH, or increasing the soil pH by adding agricultural lime will change the concentration of soluble nutrients (Havlin et al. 2005). Adding organic residues such as compost, animal manure, mulch, and crop residues will stimulate the activity of heterotrophic microorganisms, which are responsible for the biochemical transformations of organic nitrogen, phosphorus, and sulfur compounds as well as the biogeochemical cycling of micronutrients involved in oxidation-reduction reactions (e.g., iron and manganese). Irrigating soil will increase the quantity of soluble nutrients from chemical dissolution and biochemical hydrolysis.

3.3.1 Principles Guiding Fertilizer Use in Nutrient-Deficient Soils

My principle for using fertilizer in agriculture is that "Fertilizer use has to be sensible," and I consider this to be important for several reasons. First, it should make economic sense to apply fertilizer. If you spend money to buy fertilizer and pay for the labor and machinery to spread it (and consider that your own time to organize these activities is valuable), you should expect an economic return due to more crop yield or a high-quality crop that can be sold at a premium price. If there is no return on investment, then it was not economically sensible to apply fertilizer.

The second reason that "fertilizer use has to be sensible" is because the crop is the intended beneficiary of the fertilizer. We must be sensible about the way we apply fertilizer in the field, so that most of the nutrients get to the crop. This idea is fully aligned with the 4-R Nutrient Stewardship approach (IPNI 2012), which is the industry standard. Here are my suggestions according to the 4-Rs:

1. *Choose the right source of fertilizer* that will release soluble nutrients for your crop. The right source of fertilizer depends on its availability and price. If you have manure or compost on your farm, use this inexpensive, local source of nutrients first, before you purchase commercial fertilizer. If you are living more than 100 km away from the nearest livestock operation, it does not make economic sense to truck the animal manure from that operation to your farm because the breakeven distance for livestock manure is generally <30 km (Whalen et al. 2002). In this case, the right source of fertilizer for your farm would be inorganic fertilizer.

2. *Choose the right rate of fertilizer* that permits your crop to reach its yield potential. Applying more nutrients than the crop can use will not improve the yield; it will just make the weeds grow better. From an economic perspective, it is better to apply a little less fertilizer than you think you need, because applying too much fertilizer is a waste of money. Also, there is a risk associated with over-applying micronutrient fertilizers because there is a very fine line between a micronutrient deficiency and a micronutrient toxicity in crops. When selecting the right rate, consider the residual nutrients that may be provided by the decomposition of animal manure and crop residues from a previous growing season.

3. *Choose the right time to apply fertilizer* so that you can synchronize the time that soluble nutrients enter the soil pore water around the root and the most rapid phase of vegetative crop growth. Annual crops have an exponential growth phase for several weeks during their vegetative growth period, which slows down after floral initiation. The best strategy is to apply fertilizer just before the exponential growth phase, so the plants will use fertilizer nutrients with high efficiency. This also applies to perennial crops, which should be fertilized before they renew their vegetative growth each year.

4. *Choose the right place to apply fertilizer* so that you can ensure contact between soluble nutrients and the plant root. The most efficient methods for annual crops

involve placing fertilizer in the planted row (i.e., between seeds or tubers) or beside the planted row so that the roots will grow into the soil pores containing soluble nutrients. Seeds can also be treated with fertilizer, as long as the fertilizer has a low salt content so there is no osmotic stress that interferes with seed germination and early seedling development. Micronutrient fertilizers can also be applied in a liquid spray on the foliage, which allows the soluble nutrients to enter the plant through pores in the leaves.

The final reason that "fertilizer use has to be sensible" is because some of the soluble nutrients added to the soil-plant system are susceptible to loss into the environment. This is an undesirable outcome for ecological and public health reasons, described below.

3.4 Nutrient-Excessive Soils

Soils with excessive amounts of soluble nutrients are rare in nature. The soluble nutrient concentration changes with time due to biological uptake, because of chemical precipitation, occlusion, and adsorption reactions, and through hydrological pathways like leaching (via mass flow and preferential flow) and surface runoff. Ions that leave the soil pore water are gradually replenished by dissolution and mineralization processes, as well as desorption from soil organo-mineral surfaces.

There is a realization since the 1990s that many agricultural soils are approaching, or have reached, an excessively high concentration of nutrients, primarily nitrogen and phosphorus. According to Steffen et al. (2015), the amount of nitrogen and phosphorus in the Earth's biosphere is at a level that exceeds the planetary boundary for these nutrients. This is not very worrisome for plants, since they have metabolic controls that prevent them from absorbing more nitrogen and phosphorus than they can use to meet their nutrient requirements for growth and yield. However, agricultural soils have a finite capacity to retain these nutrients, which means that nitrogen and phosphorus are vulnerable to be transferred from agroecosystems to non-agricultural environments, including aquatic systems and the atmosphere.

3.4.1 Observations About Nutrient-Excessive Soils

My first exposure to soils with nutrient excesses came when I accepted a job as a Term Research Scientist with Agriculture and Agri-Food Canada in Lethbridge, Alberta, Canada. My supervisor, Dr. Chi Chang, asked me to evaluate the soil phosphorus dynamics in barley agroecosystems that had a 20+ year history of amendment with beef cattle manure. My textbook knowledge of soil phosphorus cycling was challenged immediately – according to what I knew at the time, soil had an apparently infinite capacity to absorb phosphorus. This was clearly not the case in this long-term field experiment, where the research team had applied high rates of cattle manure to a clay-loam soil, every year for more than 20 years. When I calculated the nutrient balance, I could not account for all the phosphorus that was applied to those soils. I knew the crop phosphorus removal each year, and I had the soil phosphorus concentration in the profile to a depth of 1.5 m, but there was a substantial amount of "missing" phosphorus. On this flat field, which had minimal erosion, the only explanation was that phosphorus leached in deeper soil layers and potentially had left the soil-plant system entirely – in other words, it had leached out of the soil and possibly ended up in the underlying aquifer (Whalen and Chang 2001).

After I was appointed to the position of Assistant Professor at McGill University in 2000, I continued to work on soil phosphorus biogeochemistry, among other projects. I was invited to join the provincial Soil Chemistry and Fertility Committee, which is responsible for updating the soil fertilizer recommendations for the province of Quebec, Canada, most recently the CRAAQ (2010) guidebook. Through this committee, I met agrologists who worked with the fertilizer companies, with farm groups, and with provincial ministries and research institutes. At the same time, I was teaching a course in soil fertility and fertilizers and working through local case studies with my undergraduate students. We could see clearly that soil phosphorus levels were becoming excessive on some farms with livestock, across the province.

In 2001, the Government of Quebec passed the *Agricultural Operations Regulation* legislation (CQLR 2019) that aimed to control the phosphorus balance on farms. The largest livestock producers with big flocks were required to comply with the legislation within the first 5 years, and gradually all other livestock producers, crop producers, orchards, and greenhouse operators had to follow the regulation. The basic premise of the legislation is that farmers need to test their soil and their manure regularly, so they know the fertilizer requirements of the soil and what nutrients can be provided by the manure. There are limits to the amount of phosphorus fertilizer that can be applied to agricultural fields, depending upon the soil phosphorus level and the amount of phosphorus that will be removed by crops. Corn (maize) removes more phosphorus from the field than many other crops, so fields that will be planted with corn can receive larger phosphorus inputs from manure and other fertilizers than fields that produce cereals, soybean, and hay (forage crops).

The *Agricultural Operations Regulation* was a positive step forward, since it provided clear guidelines about phosphorus management on farms and sensitized producers to the fact that they need to test their soil and manure regularly. Still, I and other scientists remained vigilant to the fact that phosphorus could be transported in agricultural fields in this area and eventually reach downstream rivers and lakes. Besides surface runoff, we were attentive to the possibility that phosphorus could be lost from fields through subsurface tile drainage systems (Enright and Madramootoo 2004). In our cold humid climate, subsurface tile drainage is often installed at a depth of 0.9–1.2 m below the soil surface. The purpose is to drain the excess water from the field, so the surface soil is dry enough to plant the crop. Subsurface drainage also allows farmers to control the level of the water table, which is important because the root system cannot grow in a flooded soil. In one study, we found that water-erodible soil particles from the topsoil were susceptible to be transported across the field surface and through the soil profile, where they exited the field in the tile drainage. Smaller particles (0.05 to 1 μm) were enriched in phosphorus, relative to coarse particles. The phosphorus concentration associated with these particles was as high as 3181 μg P L^{-1} in surface runoff and 1346 μg P L^{-1} in tile drainage (Poirier et al. 2012). Since the phosphorus level that triggers eutrophic conditions in Canadian lakes is 35–100 μg P L^{-1} (CCME 2004), it is clear that the loads from agricultural fields could be a source of phosphorus pollution in the environment. We concluded that the loss of phosphorus-rich sediments with particle size <1 μm from agricultural fields could contribute to the eutrophication of downstream water bodies.

Eutrophication of aquatic systems is a growing concern in Canada. Severe algal blooms in Lake Winnipeg and Lake Simcoe and blooms of cyanobacteria in eastern Canadian lakes have been occurring in recent years, as well as reemerging problems in Lake Ontario and Lake Erie, and in other Canadian water bodies (Government of Canada 2015). The cyanobacteria blooms are perhaps the most alarming, due to the fact that cyanobacteria produce metabolites called cyanotoxins that are lethal to aquatic organisms and any other animals that come into contact with the bloom, including birds, livestock, domestic animals, and humans.

Cyanobacteria blooms are generally associated with excessive phosphorus in the environment, but the production of some cyanotoxins appears to be associated with nitrogen excesses. For example, the cyanotoxin β-methylamino-l-alanine (BMAA) is a non-protein amino acid that is produced in response to the nitrogen availability in freshwater environments. Production of BMAA appears to provide a competitive advantage to the non-nitrogen-fixing cyanobacteria, which are competing for light, space, and other resources with the nitrogen-fixing cyanobacteria (Zhang and Whalen 2020). Lakes that receive nitrogen inputs from agricultural runoff and other sources are reported to have elevated BMAA concentrations. This is worrisome because BMAA is a suspected causative agent of human neurodegenerative diseases including Alzheimer's disease, Parkinson's disease, and amyotrophic lateral sclerosis.

Nutrient-excessive soils in many parts of the world are linked to negative outcomes in downstream water bodies. The occurrence of algal and cyanobacteria blooms is becoming more frequent worldwide. The fact that cyanobacteria produce cyanotoxins that can cause illness or death in animals and humans should encourage us to do more to control nutrient losses from agricultural fields. Nutrient management will continue to be an important topic, due to its connection to ecological and public health, for the foreseeable future.

3.5   Precision Agriculture, a Smart System for Soil Nutrient Management

Technological advances in the field of precision agriculture are a game-changer for the agricultural sector. The greatest advantage of precision agriculture, from the perspective of managing nutrients, is that it helps us to develop a spatially explicit map of the soil and hydrological conditions in agricultural fields. Such a map is created using geographic information systems, meaning it contains multiple layers of information. We can look at individual parameters, such as soil pH or the soil potassium concentration across a field, or we can evaluate the interactions between multiple parameters (e.g., are changes in soil pH associated with changes in the water table level across the field?). Rather than relying on an average soil fertility value for a large field, the farm operator can understand the gradients in soil fertility that exist across their field. Knowing the zones of low, medium, and high fertility allows the farmer to adjust fertilizer application rates to meet the needs of the crop growing in a particular zone, rather than applying a uniform rate of fertilizer across the field. Hence, the farm can save money and apply fertilizers precisely where they are needed.

A precision approach to managing nutrients should go beyond the field and can be extended to other parts of the farm operation. Consider manure, a nutrient-rich fertilizer. About 50% or more of the nitrogen in animal manure is lost to the environment, between the time that the animal defecated/urinated and the time that the manure is applied to an agriculture field. The reason that nitrogen is susceptible to loss from animal manure is because bacteria present in the fecal material produce urease, an enzyme that degrades urea in the urine. The by-product of this reaction is NH_3, a gas at room temperature. In manure-amended soil, the NH_3 produced by urea hydrolysis remains as NH_3 (aq) in the soil pore water and is rapidly protonated with H^+ ions, so it is retained in soil as the soluble NH_4^+ ion. In barns, animals generally deposit their manure onto a concrete surface. There is a high probability that NH_3 (g) released from urea hydrolysis in barns will be lost to the atmosphere. Ammonia sensors in barns will indicate when and where the volatile NH_3 (g) is being produced. There are several strategies to prevent NH_3 (g) loss. One is to dry the manure, since urea hydrolysis cannot proceed without water. This is feasible in laying hen facilities that are heated and ventilated for the comfort of the animals. Another possibility is to spray the manure with a dilute acidic solution to provide H^+ for reaction with NH_3. Some operators install filters that are impregnated with H^+ to capture the NH_3 before it leaves the barn (Whalen et al. 2019).

Preventing nutrient loss from manure piles and storages is another good practice. Manure piles should be spread on agricultural fields during the crop growing season and incorporated into the soil as soon as possible. Manure storages are made from concrete or packed earth and ideally will be covered to minimize NH_3 (g) volatilization. Covering a manure storage also prevents rain and snow from entering, which will reduce the amount of water that needs to be pumped, hauled, and spread onto the field along with the manure. Manure storages must be emptied every year. After manure is applied to the field, it needs to be incorporated into the soil as soon as possible. The best times of year to apply manure are in the spring, before planting the crop, and during the growing season. Solid and liquid manure can be top-dressed onto forages between cuts. Furthermore, solid and liquid manure can be banded between widely spaced crop rows during vegetative growth stages, when the crop nutrient demands are highest (Whalen et al. 2019). Applying manure in the fall and winter is not recommended, since nutrients are susceptible to be lost from soil during freezing-thawing events. Since nutrients are to be used for crop production, we must always apply manure when crops are growing in the field.

Manure is a valuable fertilizer that must be managed carefully and applied judiciously to get the greatest benefit in the crop production system. Technologies developed for precision agriculture should help us to recycle the maximum amount of nutrients from manure to produce more food.

3.6 Final Thoughts and Future Perspectives

I have been studying the biogeochemical cycling of nitrogen, phosphorus, and sulfur in the soil-plant system for more than 20 years. During this time, I have observed a significant shift in perceptions about soil fertility and fertilizers. The farming community understands that nutrients applied to agricultural fields are susceptible to transfer into aquatic systems and the atmosphere, with potential consequences for ecological and public health. We need fertilizers to produce food for the world, but we also need to be cautious, timely, and more sparing in our use of nutrients from all sources, including manure and residual fertilizing substances that come from households, municipalities, and industries. There will be more potential fertilizers available in the future because it is no longer socially acceptable to landfill organic wastes and other nutritive substances. Residual fertilizing substances contain organic carbon and nutrients that should be returned to farms, forests, and rangelands. The dilemma is that residual fertilizing substances often contain organic and inorganic pollutants, as well as emerging contaminants such as microplastics, nanoparticles, antibiotics, and hormones. Research is needed on the safe landspreading techniques for residual fertilizing substances, so we can benefit from their nutritive value while avoiding negative ecological impacts.

In the future, soil fertility researchers and practitioners can explore alternative ways to boost soil fertility through the use of biostimulants. These are substances(s) and/or microorganisms whose function, when applied to plants or the rhizosphere, is to stimulate natural processes to enhance or benefit nutrient uptake, nutrient efficiency, tolerance to abiotic stress, and crop quality (EBIC 2012). I always had the sense that soil biology makes a meaningful contribution to plant nutrition, based on my own experience with earthworms and soil microbial communities. Biostimulants enhance the capacity of soil biota to provide multiple, positive benefits to the plants. Stimulating decomposition and mineralization processes is one way that biostimulants increase nutrient availability. Some biostimulants contain chelating substances that bind ions present in the soil solution and keep them in a form that is easily adsorbed through the plant roots. There is the possibility that biostimulants can help crops to acquire more nutrients from the soil environment. Improving nutrient use efficiency in the soil-plant system with biostimulants has the potential to reduce the amount of fertilizer required for crop production.

Biostimulants have additional modes of action that are positive for crops. The production of auxin-like substances and other plant growth hormones is another way that some biostimulants contribute to the proliferation of an extensive root system with many lateral branches and fine roots. On foliage, plant growth hormones promote the growth of apical meristems, resulting in more stems and leaves. Some biostimulants enhance plant defenses by upregulating the jasmonate or salicylate defense pathways. Biostimulants that contain endophytic microorganisms may protect plants by colonizing the surface of roots, stems, and leaves, thereby creating a physical barrier to disease-causing organisms. Some humic substances and seaweed extracts contain betaines, organic osmolites that help to maintain the homeostasis in plant cells and prevent osmotic stress in plants exposed to drought or salt stress.

While it is unlikely that any one biostimulant can perform all of the functions mentioned above, it is encouraging that we can rely upon biological processes and naturally produced substances to support greater resilience in the soil-plant system. Instead of simply managing the soil nutrients, we can now envision innovative ways to sustain the resilience and regenerative power of the living soil around every plant, to ensure an abundant, nutritious food supply, now and for future generations.

References

Canadian Council of Ministers of the Environment (CCME). (2004). Canadian water quality guidelines for the protection of aquatic life: Phosphorus: Canadian guidance framework for the management of freshwater systems. In *Canadian environmental quality guidelines, 2004.* Winnipeg, Canada: Canadian Council of Ministers of the Environment.

Centre de référence en agriculture et agroalimentaire du Québec (CRAAQ). (2010). *Guide de référence en fertilisation.* 2e édn. Ste Foy, Canada: Centre de référence en agriculture et agroalimentaire du Québec.

Compilation of Québec Laws and Regulations (CQLR) §c Q-2, r 26 (2019). *Agricultural operations regulation.* http://canlii.ca/t/52z8z

Enright, P., & Madramootoo, C. A. (2004). Phosphorus losses in surface runoff and subsurface drainage waters on two agricultural fields in Quebec. In R. A. Cooke (Ed.), *Drainage: Proceedings of the 8th international symposium, Sacramento, CA, 21–24 Mar 2004* (pp. 160–170). St. Joseph, MI: American Society of Agricultural and Engineering.

European Biostimulants Industry Council (EBIC). (2012). *What are biostimulants?* http://www.biostimulants.eu/about/what-are-biostimulants

Government of Canada. (2015). *Phosphorus in aquatic systems.* https://www.canada.ca/en/environment-climate-change/services/freshwater-quality-monitoring/publications/phosphorus-aquatic-ecosystems.html

Havlin, J. L., Beaton, J. D., Tisdale, S. L., & Nelson, W. L. (2005). *Soil fertility and fertilizers: An introduction to nutrient management* (7th ed.). Upper Saddle River: Pearson Prentice Hall.

Hubbs, T. (2019). Weekly outlook: acreage in 2020. *farmdoc daily* (9), 222. Department of Agric Consumer Econ, Univ Illinois at Urbana-Champaign, 25 Nov 2019. https://farmdocdaily.illinois.edu/2019/11/weekly-outlook-acreage-in-2020.html

International Plant Nutrition Institute (IPNI). (2012). *4R plant nutrition: A manual for improving the management of plant nutrition*. Norcross: North American Version, IPNI.

Jenny, H. (1980). *The soil resource – Origin and behavior* (Ecological studies) (Vol. 37). New York: Springer.

Jones, J. B., Jr. (2003). *Agronomic handbook: Management of crops, soils and their fertility*. Boca Raton: CRC Press.

Poirier, S.-C., Michaud, A. R., & Whalen, J. K. (2012). Bioavailable phosphorus in fine-sized sediments transported from agricultural fields. *Soil Science Society of America Journal, 76*, 258–267.

Reid, R. S., Fernández-Giménez, M. E., & Galvin, K. A. (2014). Dynamics and resilience of rangelands and pastoral peoples around the globe. *Annual Review of Environment and Resources, 39*, 217–242.

Sims, J. T., Vadas, P. A. (1997). *Use of phytase and low-phytate corn to increase phosphorus availability in poultry feed*. Fact Sheet ST-10, College Agric Sci Coop Ext, Univ Delaware, Newark, DE.

Statista. (2020). *Grain production worldwide 2018/19, by type*. https://www.statista.com/statistics/263977/world-grain-production-by-type/

Steffen, W., Richardson, K., Rockström, J., Cornell, S. E., et al. (2015). Planetary boundaries: Guiding human development on a changing planet. *Science, 347*(736), 1259855.

United Nations Department of Economic and Social Affairs (UN DESA). (2019). *World population prospects 2019*. https://population.un.org/wpp/

Whalen, J. K., & Chang, C. (2001). Phosphorus accumulation in cultivated soils from long-term annual applications of cattle feedlot manure. *Journal of Environmental Quality, 30*, 229–237.

Whalen, J. K., & Parmelee, R. W. (1999). Quantification of nitrogen assimilation efficiencies and their use to estimate organic matter consumption by the earthworms *Aporrectodea tuberculata* (Eisen) and *Lumbricus terrestris* L. *Applied Soil Ecology, 13*, 199–208.

Whalen, J. K., & Warman, P. R. (1996). Arylsulfatase activity in soil and soil extracts using natural and artificial substrates. *Biology and Fertility of Soils, 22*, 373–378.

Whalen, J. K., Parmelee, R. W., McCartney, D. M., & VanArsdale, J. L. (1999). Movement of N from decomposing earthworm tissue to soil, microbial and plant N pools. *Soil Biology and Biochemistry, 31*, 487–492.

Whalen, J. K., Bottomley, P. J., & Myrold, D. D. (2000). Carbon and nitrogen mineralization from light- and heavy-fraction additions to soil. *Soil Biology and Biochemistry, 32*, 1345–1352.

Whalen, J. K., Chang, C., & Clayton, G. W. (2002). Cattle manure and lime amendments to improve crop production of acidic soils in Northern Alberta. *Canadian Journal of Soil Science, 82*, 227–238.

Whalen, J. K., Thomas, B. W., & Sharifi, M. (2019). Novel practices and smart technologies to maximize the nitrogen fertilizer value of manure for crop production in cold humid temperate regions. *Advances in Agronomy, 153*, 1–85.

Zhang, Y. Y., & Whalen, J. K. (2020). Production of the neurotoxin beta-N-methylamino-L-alanine may be triggered by agricultural nutrients: An emerging public health issue. *Water Research, 170*, 115335. https://doi.org/10.1016/j.watres.2019.115335.

Dr. Joann K. Whalen is a James McGill Professor at McGill University and an Adjunct Professor at Gansu Agricultural University and with the Northeast Institute of Geography and Agroecology, Chinese Academy of Sciences. She received her PhD from Ohio State University (USA) and worked as a research scientist for Agriculture and Agri-Food Canada prior to joining the faculty at McGill. Dr. Whalen is also a professional agronomist (agronome) in Quebec, Canada. Her research focuses on soil fertility and soil ecology in agroecosystems. She has published more than 200 peer-reviewed scientific publications and supervised/ co-supervised more than 70 students at the M.Sc., Ph.D., and postdoctoral levels. She teaches courses in soil fertility, nutrient management planning, and environmental soil chemistry. Dr. Whalen is a Chief Editor for Soil Biology and Biochemistry, a Subject Editor for Applied Energy and a member of the Editorial Board for Nutrient Cycling in Agroecosystems. She is the first author of the textbook *Soil Ecology and Management*, published in 2010 by CABI Publishers and the editor of *Soil Fertility Improvement and Integrated Nutrient Management: A Global Perspective*, an online book published in 2012 that presents 15 invited chapters written by leading soil fertility experts from more than 20 countries.

Chapter 4
Toward Improved Nitrogen Fertilization with Precision Farming Based on Sensor and Satellite Technologies

Heide Spiegel, Taru Sandén, Laura Essl, and Francesco Vuolo

Contents

4.1 Introduction

Nitrogen (N) is needed for all forms of life and is often the most limiting and crucial plant nutrient in European agricultural fields. Although N_2 is the most abundant element in the atmosphere, it cannot be used by most of the plants and has to be converted into reactive nitrogen (Nr). Only few plants are able to do this with the help of bacteria (biological N fixation). For most other crops, N, which is mainly stored in organic forms in the soil, must be mineralized into Nr by soil microorganisms. These microorganisms, however, also need N for energy, making them competitors with plants for Nr. Microorganisms are often initially more successful, and N is stored in their biomass. This N is released only after their necromass is mineralized (Liang et al. 2017). The energy-intensive Haber-Bosch process (Fowler et al. 2013) enables Nr to be applied as mineral fertilizer, providing crops with plant-available

H. Spiegel (✉) · T. Sandén
Department for Soil Health and Plant Nutrition, Austrian Agency for Health and Food Safety, Vienna, Austria
e-mail: adelheid.spiegel@ages.at

L. Essl · F. Vuolo
Institute of Geomatics, University of Natural Resources and Life Sciences, Vienna (BOKU), Vienna, Austria

© Springer Nature Switzerland AG 2021
T. K. Hamrita (ed.), *Women in Precision Agriculture*, Women in Engineering and Science, https://doi.org/10.1007/978-3-030-49244-1_4

nitrate and/or ammonium. Nonetheless, if nitrate is not taken up by the crop, it tends to leach into ground or surface water, causing adverse environmental effects. In principle, sustainable nutrient management should involve fertilization that replaces the amount of nutrients removed by crops (Frossard et al. 2009). On one hand, N deficiency may cause poor yields and low product quality, on the other, Nr provided in excess causes environmentally harmful nitrate leaching and eutrophication (Hawkesford 2014; Spiegel et al. 2010). Finally, N losses to the atmosphere through ammonia (NH_3) volatilization, along with emissions of the primary and secondary greenhouse gases nitrous oxide (N_2O) and NO, are becoming increasingly important climate change factors (Olfs et al. 2005; Butterbach-Bahl 2011; Butterbach-Bahl et al. 2013). N_2O and NH_3 emissions can be reduced through optimal farm management and by improving nitrogen use efficiency (Jarvis et al. 2011). The European Union has introduced the Nitrate Directive (COUNCIL DIRECTIVE 91/676/EEC) and national nitrate action programs to regulate the application of nitrogen. This is designed to yield an integrated approach with a greater emphasis on the protection of humans and nature.

Agricultural field experiments are proven tools to test and evaluate optimal N fertilization amounts for specific crops under different soil and climate conditions. When only N fertilization is varied, field experiments provide valid observations for researchers and practitioners to study the effects of low, medium, and high fertilization amounts on plant growth. New technologies such as sensor and satellite data can help farmers to find optimum N fertilization rates with a high nitrogen use efficiency (NUE), improving yields with a minimum of N losses. In the present study, remote sensing experts, soil scientists, and agronomists worked closely together to apply new technologies for improving N management. As a first step, we used long-term small test fields. We then established large-scale experimental fields in the Marchfeld in Lower Austria and compared field and satellite observations. In this area, NO_3 groundwater concentrations are often above the limit (50 mg NO_3-N l^{-1}), making it a prospective area for mitigation of groundwater pollution. As a final step, a technical application was developed to test the site-specific fertilization and to initiate the transition from theory to practice. This book chapter introduces new tools and techniques and shares the experience gained in making them applicable for farmers.

4.2 Field Experiments and Technologies

4.2.1 Field Experiments

Cross-site agricultural field experiments with different fertilization stages have been carried out for decades to derive optimal crop-specific fertilization amounts, including nitrogen (N). Long-term field experiments with distinct soil nutrient inputs in the different variants are also useful to test new versus established technologies. In contrast, experiments with small plots are better suited to test new field sensor technologies rather than satellite (Copernicus Sentinel-2) data. Accordingly, in 2015, we used two long-term field experiments from the Austrian Agency for Health and Food Safety (AGES) with different nutrient N stages to evaluate established (Nmin sampling, N tester (SPAD)) and new sensor (N-Pilot©) techniques to optimize N fertilization for winter wheat. The sites are located in Breitstetten (48.196431, 16.709064) and Haringsee (48.195190, 16.769662). Four different N fertilization stages (70, 95, 120, 145 kg N ha^{-1}) and P and K fertilization at optimum were used (3 replicates each). In Breitstetten, the variety "Energo," and in Haringsee, the variety "Midas" (both quality wheat varieties) were sown. N concentrations (%) were calculated to "raw protein content" using the factor 5.7.

In the growing season 2015/2016, we implemented large-scale field experiments with winter wheat and maize at two neighboring fields near Engelhartstetten in the Marchfeld (48°11'N 16°55'E); in 2016/2017, winter wheat was grown in the field after maize. The large plots enabled a validation of satellite data with a resolution of 10 m × 10 m. The field experiments consisted of a randomized block design with three replicates, the plots measuring 100 m × 100 m (1 ha) each. In addition to the zero fertilization variant N0 (0 kg N), the winter wheat trials consisted of three different fertilization levels with the following total fertilization quantities (N1: 60 kg N; N2: 120 kg N; N3: 180 kg N). These were applied as calcium ammonium nitrate and distributed evenly over three fertilization dates. Table 4.1 shows the site characteristics of the Marchfeld, and Table 4.2 shows the management data in the growing season from 2014/2015 to 2016/2017.

An example for the design of the experiments (winter wheat 2016) in Engelhartstetten is given in Fig. 4.1. Phosphorus and potassium fertilization was done according to the Austrian guidelines for fertilization (BMLFUW 2017). The management data of the growing seasons 2015/2016 and 2016/2017 are shown in Table 4.2.

In addition to validating the satellite data, the field test served as a basis for evaluating the efficiency of the different fertilizer quantities.

Table 4.1 Description of the experimental site Marchfeld (Breitstetten, Haringsee, Engelhartstetten)

	Unit	Marchfeld
Meters above sea level	m	147
Mean annual rainfall (∅1981–2010)	mm	516
Mean annual temperature (∅1981–2010)	°C	10.3
Soil type (IUSS 2015)		Calcaric chernozem
Texture (sand/silt/clay)	%	(17/49/34)
pH CaCl$_2$		7.54
Carbonate content	%	23.2
Soil organic carbon	%	2.36
N$_{total}$	%	0.24

Table 4.2 Management data for growing seasons 2014/2015, 2015/2016, and 2016/2017 (winter wheat)

Cropping season	2014/2015	2014/2015	2015/2016	2016/2017
Site	Breitstetten	Haringsee	Engelhartstetten	Engelhartstetten
Tillage	November 2014	November 2014	November 2015	November 9 and 11, 2016
Drilling	November 2014	November 2014	November 2015	November 15, 2016
1st N fertilization	March 2015	March 2015	March 24, 2016	March 16, 2017
2nd N fertilization	April 2015	April 2015	April 21, 2016	April 25, 2017
3rd N fertilization	May 2015	May 2015	May 19, 2016	May 30, 2017
Harvest	July 2015	July 2015	July 26, 2016	July 20, 2017

Fig. 4.1 Scheme of the fertilization experiment at Engelhartstetten (Marchfeld, Austria) in 2016. Background image: Copernicus Sentinel-2A Leaf Area Index (LAI) map acquired on May 6, 2016 (Vuolo et al. 2019)

All results are indicated as arithmetic means (n = 3). Statistical analyses involved multiple analysis of variance and subsequent multiple range tests (Tukey's honest significant difference). Crop and sensor data were correlated with bivariate correlation statistics according to Spearman. All calculations were performed using the SPSS package 22.

4.2.2   Technologies

For practical application, Copernicus Sentinel-2 data were used to calculate the Leaf Area Index (LAI) during the entire vegetation period. The LAI correlates with the existing biomass, and a combination of the observations from a large number of satellite images at different points in time correlates with crop productivity under the given growth conditions. Based on the LAI, an indicator for potential productivity has been developed and optimized for nitrogen fertilization (Novelli et al. 2019; Olfs et al. 2005). In addition, an online platform for the automatic generation of management zones was developed with automatic segmentation. Using this platform, users can determine the respective quantities of nitrogen fertilizer and individually define the desired fertilizer application parameters. The satellite data and the LAI maps were taken from the Copernicus Sentinel-2 portal of the Institute of Geomatics, University of Natural Resources and Life Sciences, Vienna (BOKU) (Vuolo et al. 2016). The processed data are automatically stored in a database; both individual images and the productivity map are available for creating the N application maps. Depending on the time of use, current or historical maps can be used to optimize N fertilization. The following sections describe the steps to derive and evaluate these technologies.

4.3   Monitoring of Biomass Production (and N Parameters in Soil and Grain)

Tables 4.3 and 4.4 show the results of conventional soil and plant measurements in two field experiments designed to achieve and evaluate optimal winter wheat yields and quality with minimum N losses to the environment during the growing season (Breitstetten site, Table 4.3; Haringsee site, Table 4.4).

In late May, the N-Pilot © revealed significant differences between 70 kg N and 145 kg N fertilization at both sites. Only in Breitstetten could the N tester distinguish between 70,120 and 145 kg N fertilization stages. This was also the case for the yield results. In Haringsee, yields were significantly higher with 120 and

Table 4.3 Effects of different N fertilization amounts on N tester (27/05/2015) and N-Pilot © measurements, winter wheat yields, N concentrations, and N uptake by the winter wheat grain at Breitstetten. Different letters: significant differences between treatments (p < 0.05; Tukey)

N fertilization kg ha^{-1} year^{-1}	N tester 27/05/15		N-Pilot © 27/05/15		Grain yield (kg ha^{-1}) 18/7/2015)		N% grain	N uptake kg ha^{-1}	
70	589	a	815	a	5113	a	2.32	102	a
95	615	ab	817	ab	5455	ab	2.31	108	a
120	649	bc	827	ab	5837	b	2.48	124	b
145	661	c	837	b	6604	c	2.58	147	c

Table 4.4 Effects of different N fertilization amounts on N tester (27/05/2015) and N-Pilot ©
measurements, winter wheat yields, N concentrations, and N uptake by the winter wheat grain at
Haringsee. Different letters: significant differences between treatments (p < 0.05; Tukey)

N fertilization kg ha^{-1} year^{-1}	N tester 27/05/15		N-Pilot © 27/05/15		Grain yield (kg ha^{-1}) 18/7/2015)		N% grain	N uptake kg ha^{-1}	
70	677	a	847	a	6401	a	2.13	117	a
95	657	a	852	ab	6847	ab	2.12	125	ab
120	703	a	860	ab	6931	b	2.27	135	ab
145	699	a	863	b	7228	b	2.37	148	b

145 kg N compared to 70 kg N fertilization. Fertilization stages of 120 and 145 kg N
yielded significantly higher raw protein (=N% × 5.7) contents in Breitstetten
(according to the Austrian "quality wheat" level: 14% RP) compared to the lower
stage (milling wheat quality >12.5% RP). In Haringsee, N milling wheat quality
was achieved with the two highest fertilization amounts. As noted above, the grain
N concentration was (significantly) higher with higher N fertilization (N3, N4).

In Breitstetten, the correlations were significantly positive between final winter
wheat grain yields, N concentrations in the grain, and sensor (N-Pilot© and N tester) measurements in late May (Table 4.5). In Haringsee, only the N-Pilot© measurements correlated positively with grain N concentrations. Furthermore,
ground-based LAI measurements were made at Breitstetten (May 11, 2015) and
corresponded well with the different fertilization stages (Fig. 4.2). The agreement
between ground-based and satellite-based (Copernicus Sentinel-2) LAI measurement was also good on larger-scale measurements in the same area (Vuolo et al.
2017, 2019).

Based on the satisfactory results of small plot experiments, we established and
conducted large-scale experiments in Engelhartstetten during the vegetation period
between 2015/2016 and 2016/2017 in order to test satellite measurements (Sentinel
−2 LAI maps) as well (Fig. 4.3). LAI satellite estimates were validated using ground
estimations, and the results showed a very good agreement (Vuolo et al. 2017).

Furthermore, the satellite-derived LAI in April (7 and 28) and on May 17, 2016
and the winter wheat grain yields were positively correlated. This was also the case
in the following year in May (3 and 28) 2017 (Fig. 4.4).

Field experiments showed that both well-established and new technologies are
available to detect the site-specific development of the vegetation. The first group
included N tester measurements; the second N-Pilot© (BOREALIS) and LAI
ground- and satellite (Copernicus Sentinel-2)-derived data. All revealed high positive correlations with the winter wheat yields. The next step was to use satellite
(Copernicus Sentinel-2A/B) data, which provide images with high spectral and spatial resolution with 13 spectral bands of 10–60 m spatial resolution covering the
whole earth every 5 days (Vuolo et al. 2016). These Sentinel-2 surface reflectance
data serve as a basis to produce potential productivity maps and N application maps
(see Sect. 4.4).

Table 4.5 Correlations (Spearman) between sensor (N-Pilot © and N tester on 27,052,015) measurements, grain yields, and N concentrations of winter wheat in Breitstetten and Haringsee

	Breitstetten		Haringsee	
	Winter wheat grain yield dt ha^{-1}	N% wheat grain	Winter wheat grain yield dt ha^{-1}	N% wheat grain
N-Pilot©	0.49**	0.62*	0.24 n.s.	0.69*
27/05/15	$n = 36$	$n = 12$	$n = 36$	$n = 12$
N tester	0.82**	0.70*	0.46 n.s.	0.41
27/05/15	$n = 12$	$n = 12$	$n = 12$	$n = 12$

**significant at $P < 0.01$, *significant at $P < 0.05$

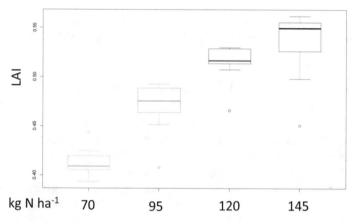

Fig. 4.2 First results of LAI ground-based measurements (BOKU) on May 11, 2015 at Breitstetten

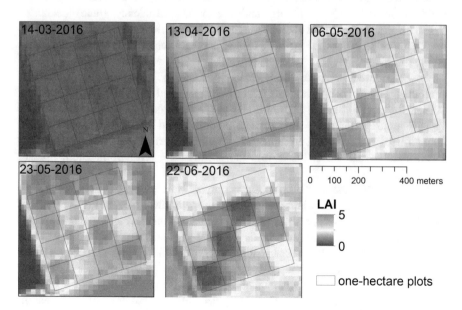

Fig. 4.3 Series of LAI maps of the field experiment with winter wheat in Engelhartstetten 2016, indicating the different N fertilization amounts already in the LAI map derived in April (Vuolo et al. 2017)

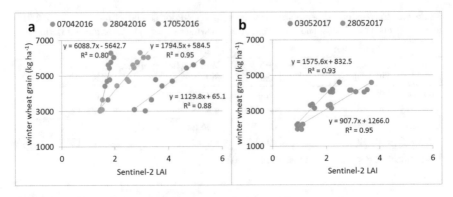

Fig. 4.4 Relationship between the winter wheat grain yields and Copernicus Sentinel-2 LAI derived (**a**) in April (7 and 28) and on May 17, 2016 and (**b**) in May (3 and 28) 2017

4.4 Ecological Aspects (Nitrogen Use Efficiency, Nmin)

Optimal N fertilization is characterized by a high N use efficiency and is crucial to minimizing N losses to water and air. One objective of sustainable N management is to ensure that a maximum share of the fertilized nitrogen is taken up by plant roots, is optimally used in the plant for protein formation, and is then available within the harvested crop for human use. A meaningful OECD agri-environmental indicator is based on an N output/N input analysis (N removal/N fertilization) (OECD 2019; Brentrup and Palliere 2010; Johnston and Poulton 2009). Figure 4.5 shows a well-known effect at Engelhartstetten: With increasing N fertilization, the yield curve (orange) levels off, and the N use efficiency (blue) decreases (Hawkesford, 2014; Olfs et al. 2005). According to Vuolo et al. (2019), the N fertilization increase from 120 kg ha^{-1} to 180 kg ha^{-1} resulted in a negligible economic return.

An often used indicator for potential N leaching losses during postharvest autumn and winter is the Nmin soil test (Wehrmann and Scharpf 1979; ÖNORM L 1091). This method helps determine the plant-available mineral nitrogen (nitrate and ammonium N) in the rooting zone (0–90 cm). When N fertilization rates exceed the optimum, the residual soil mineral N after harvest reportedly increases, posing a considerable risk for N losses into the groundwater (Olfs et al. 2005; Spiegel et al. 2010).

Nmin after harvest in 2015 (measured only in Haringsee) was twice as high in the 145 kg fertilization variant compared to the two lower N stages (Fig. 4.6a). In Engelhartstetten as well, postharvest Nmin measurements revealed that this parameter was almost twice as high in the highest versus lower fertilization stages (Fig. 4.6c). Thus, two of our experiments confirm that fertilization exceeding a certain optimum application rate (95 kg N ha^{-1} year^{-1} in Haringsee 2015 and 120 kg N ha^{-1} year^{-1} in Engelhartstetten 2017) results in an increase of the residual soil Nmin after harvest. This, however, was not the case for winter wheat in

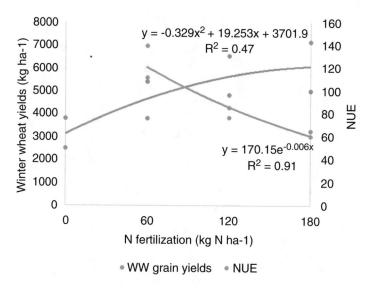

$y = -0.329x^2 + 19.253x + 3701.9$

$R^2 = 0.47$

$y = 170.15e^{-0.006x}$

$R^2 = 0.91$

• WW grain yields   • NUE

Fig. 4.5 Winter wheat grain yields and NUE (N use efficiency: N output/N fertilization) in Engelhartstetten in 2016 and 2017

Engelhartstetten in 2016 (Fig. 4.6b). One explanation is that the applied fertilizer N was utilized and transferred into biomass production because of sufficient and well-distributed precipitation patterns over the vegetation period. In 2017, drought periods decreased winter wheat yields and, thus, plant N uptake. The N not used by the crop is traceable as Nmin (especially nitrate) in the soil and is highly prone to being lost by leaching during autumn and winter.

Both weather conditions and soil conditions within a field influence N uptakes by crops and influence potential N losses into the groundwater. Especially in the Marchfeld study area, soil conditions are spatially highly variable. Sandy, shallow soils in which the gravel body rises high alternate with deeper, loamier soils that can store more nitrogen and provide different crop growth conditions. A site-specific adaptation of N fertilization (variable-rate application) according to crop needs would help avoid over- or underfertilization, improve yields, and minimize adverse environmental N losses (Olfs et al. 2005).

4.5   Use of Technical Devices in Agricultural Practice

For approximately 20 years, tools for precision farming have been developed to better manage agricultural fields according to the specific spatial differences in soil and crop parameters. To improve N fertilization, optical sensors can now measure canopy reflectance and enable drawing conclusions about a crop's biomass and N status. Here, the greenness (proxy for the chlorophyll content) is taken to estimate the

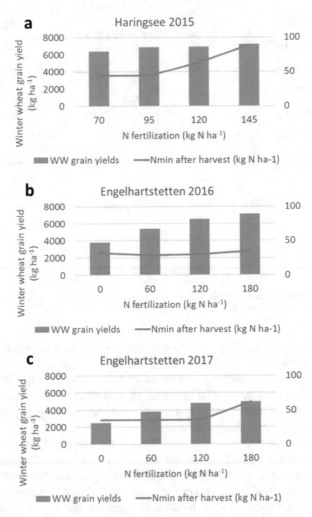

Fig. 4.6 Winter wheat yields and soil mineral N (Nmin) after harvest in (**a**) Haringsee 2015, (**b**) Engelhartstetten 2016, and (**c**) Engelhartstetten 2017

leaf N. Handheld devices such as the chlorophyll meter SPAD (N tester) have been successfully employed by using a relative approach comparing an overfertilized plot with the relevant field (Olfs et al. 2005). Another method is to correct SPAD measurements based on the variety's specific greenness, factor in the current N supply from fertilization and mineralization, and then determine the further plant N need. This same goal was pursued by BOREALIS, which recently developed a device (N-Pilot©) that was easier to handle. At the same time, that device required a comparison with a very well-fertilized canopy to derive the actual N status and the crop's further fertilization demand. Our earlier and current field experiments show that both devices yield a positive correlation with crop yields and were helpful in deriving additional fertilization needs for winter wheat. This was especially useful to dose the last (quality) fertilizer input to improve raw protein content in the winter wheat grain, which improves the economic return.

Since the launch of the Copernicus Sentinel-2 satellites, spectral information at high spatial resolution every 5 days (since 2017) can be used for agricultural purposes (Vuolo et al. 2016). Here, we used the results of the Engelhartstetten field experiments in 2016 and 2017 to develop vegetation indices to derive the canopy N status. These then served to validate the satellite data. In a subsequent step, productivity maps were created. The underlying calculations integrate plant development over several years and the local climate. This approach helps derive N application zones (Fig. 4.7). The productivity map can be made available to users in an online application (Fig. 4.8). The user can choose between the most recent satellite image and the productivity map based on historical images to define the zones for site-specific N fertilization (Vuolo et al. 2019). The necessary input parameters are the minimum area of the individual zones and the lower and upper limits of the nitrogen application. The latter limits are derived from the Austrian guidelines for appropriate fertilization (BMLFUW 2017). Importantly, the fertilizer strategy must be selected. The farmer may choose between "catch up," i.e., provide more where biomass is less, and "top down," i.e., provide less where biomass is less. For nitrate-vulnerable zones, the second strategy is preferable to avoid N leaching. For winter cereals, the first application often takes place in early March, before the spatial variability of the canopy is visible. In this case, the productivity map is useful to create the N application map. Information on the preceding crop, actual weather conditions, and Nmin contents should also be taken into account. For the second

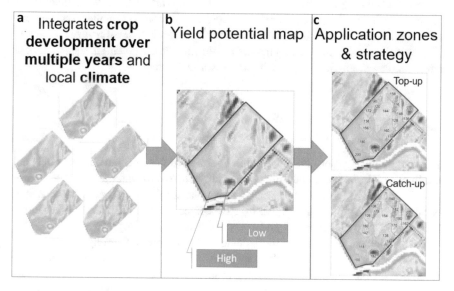

Fig. 4.7 Observation of crop development over several years and the local climate (**a**), development of potential yield maps (**b**), and derivation of application zones (**c**)

Fig. 4.8 Screenshot of the online tool for the generation of N application maps for variable-rate fertilization

and third fertilization, current satellite images help respond to the current canopy conditions. In a final step, the end user can download the data in the required format and then transfer them directly to the tractor terminal for variable-rate fertilization. This tool was successfully validated on two practice areas (Vuolo et al. 2019). It was implemented in an agricultural management and decision tool ("Agrarcommander"), which is widely used by farmers. The advantage of the system is that it can be used for smaller fields ($\geq$1 ha) and can be operated with standard agricultural machinery. Even if the tractor is not equipped with GPS and onboard computer, farmers can load the maps onto their mobile devices and apply more or less fertilizer manually as they drive across the field. This application has great potential, especially for small- and medium-sized farms without sensor technology.

4.6 Conclusions

Efficient N fertilization, one that helps optimize yields and farmer revenues and minimize environmental and climate-relevant N losses to water and air, is a major global challenge. For many years, sensor technologies (e.g., SPAD meter, N-Sensor™) and soil tests (e.g., Nmin) have helped to adapt nitrogen fertilization to the specific crop development and soil conditions and to improve the N use efficiency. Nonetheless, the use of these devices can be time consuming and/or expensive. New satellite (Copernicus Sentinel-2 data) data with a good spectral and spatial resolution enable creating and providing productivity maps to derive N variable-rate application zones. Deriving optimum N fertilizer quantities also calls for incorporating information on the preceding crop, actual weather conditions, and Nmin contents. In the near future, validation activities will be expanded here and else-where, and farmers will become increasingly better acquainted with the use of sat-ellite-derived application maps.

Acknowledgments Financial support was provided by the "FATIMA" Project. This project has received funding from the European Union's Horizon 2020 research and innovation program under grant agreement No 633945.

References

BMLFUW. (2017). Richtlinien für die Sachgerechte Düngung (2017) *Anleitung zur Interpretation von Bodenuntersuchungsergebnissen in der Landwirtschaft*. 7. Auflage. Wien. Bundesministerium für Land- und Forstwirtschaft, Umwelt und Wasserwirtschaft.

Brentrup, F., & Palliere, C. (2010, March 23–26). Nitrogen use efficiency as an agro- environmental indicator. In: *Conference: OECD Workshop on Agri-Environmental Indicators*. Leysin, Switzerland.

Butterbach-Bahl, K. (2011). Nitrogen processes in terrestrial ecosystems. In A. Bleeker, B. Grizzetti, C. M. Howard, G. Billen, H. van Grinsven, J. W. Erisman, M. A. Sutton, & P. Grennfelt (Eds.), *The European nitrogen assessment: Sources, effects and policy perspectives* (pp. 99–125). Cambridge: Cambridge University Press.

Butterbach-Bahl, K., Baggs, E. M., Dannenmann, M., Kiese, R., & Zechmeister-Boltenstern, S. (2013). Nitrous oxide emissions from soils: How well do we understand the processes and their controls? *Philosophical Transactions of the Royal Society of London. Series B, Biological Sciences, 368*, 20130122.

Fowler, D., Coyle, M., Skiba, U., Sutton, M. A., Cape, J. N., Reis, S., Sheppard, L. J., Jenkins, A., Grizzetti, B., Galloway, J. N., Vitousek, P., Leach, A., Bouwman, A. F., Butterbach-Bahl, K., Dentener, F., Stevenson, D., Amann, M., & Voss, M. (2013). The global nitrogen cycle in the twenty-first century. *Philosophical Transactions of the Royal Society, B: Biological Sciences, 368*, 20130164.

Frossard, E., Bünemann, E., Jansa, J., Oberson, A., & Feller, C. (2009). Concepts and practices of nutrient management in agro-ecosystems: Can we draw lessons from history to design future sustainable agricultural production systems? *Bodenkultur, 60*, 43–60.

Hawkesford, M. J. (2014). Reducing the reliance on nitrogen fertilizer for wheat production. *Journal of Cereal Science, 59*, 276–283.

IUSS Working Group WRB. (2015). *World reference base for soil resources 2014, update 2015. International soil classification system for naming soils and creating legends for soil maps* (World soil resources reports no. 106). Rome: FAO.

Jarvis, S., Hutchings, N., Brentrup, F., Olesen, J. E., van de Hoek, K. W., Sutton, M. A., Howard, C. M., Erisman, J. W., Billen, G., Bleeker, A., Grennfelt, P., van Grinsven, H., & Grizzetti, B. (2011). Nitrogen flows in farming systems across Europe. In *The European nitrogen assessment* (pp. 211–228). Cambridge: Cambridge University Press.

Johnston, A. E., & Poulton, P. R. (2009). *Nitrogen in agriculture: An overview and definitions of nitrogen use efficiency*. York: International Fertiliser Society.

Liang, C., Schimel, J. P., & Jastrow, J. D. (2017). The importance of anabolism in microbial control over soil carbon storage. *Nature Microbiology, 2*, 17105.

Novelli, F., Spiegel, H., Sandén, T., & Vuolo, F. (2019). Assimilation of Sentinel-2 leaf area index data into a physically-based crop growth model for yield estimation. *Agronomy, 9*(5), 255. https://doi.org/10.3390/agronomy9050255.

OECD. (2019). *Trends and drivers of agri-environmental performance in OECD countries*. Paris: OECD.

Olfs, H.-W., Blankenau, K., Brentrup, F., Jasper, J., Link, A., & Lammel, J. (2005). Soil- and plant-based nitrogen-fertilizer recommendations in arable farming. *Journal of Plant Nutrition and Soil Science, 168*, 414–431.

Spiegel, H., Dersch, G., Baumgarten, A., & Hösch, J. (2010). The International Organic Nitrogen Long-term Fertilisation Experiment (IOSDV) at Vienna after 21 years. *Archives of Agronomy and Soil Science, 56*, 405–420.

Vuolo, F., Żółtak, M., Pipitone, C., Zappa, L., Wenng, H., Immitzer, M., Weiss, M., Baret, F., & Atzberger, C. (2016). Data service platform for Sentinel-2 surface reflectance and value-added products: System use and examples. *Remote Sensing, 8*, 938.

Vuolo, F., Essl, L., Zappa, L., Sandén, T., & Spiegel, H. (2017). Water and nutrient management: The Austria case study of the FATIMA H2020 project. *Advances in Animal Biosciences, 8*, 400–405.

Vuolo, F., Essl, L., Sandén, T., & Spiegel, A. (2019). Multidisziplinäre Überlegungen zur nach-haltigen N-Düngung unter Berücksichtigung der Möglichkeiten der satellitengestützten Präzisionslandwirtschaft . Dreiländertagung OVG – DGPF – SGPF, Vienna, Feb 20–22, 2019.In: Kersten TP (Hrsg.), *Dreiländertagung OVG – DGPF – SGPF Photogrammetrie - Fernerkundung - Geoinformation – 2019*. ISBN: 0942–2870.

Wehrmann, J., & Scharpf, H. C. (1979). Der Mineralstickstoffgehalt des Bodens als Maßstab für den Stickstoffdüngerbedarf (Nmin-Methode). *Plant and Soil, 52*, 109–126.

Dr. Heide Spiegel is a Senior Scientist at the Austrian Agency for Health and Food Safety (AGES), Austria. She is a docent for soil ecology at the University of Natural Resources and Life Sciences (BOKU Vienna), where she lectures on the subject "Soils and Global Change." She is monitoring the effects of different soil management on soil properties related to nutrient (especially nitrogen and phosphorus) and organic matter cycling as well as soil contamination (acidification, enrichment with selected elements). She is the scientific manager of (long-term) field experiments and participates in national and international research projects.

Dr. Taru Sandén received her doctoral degree in Geography from the University of Iceland in 2014, in which she focused on soil aggregates and soil organic matter in European agricultural soils. Currently she works as Senior Expert at the Austrian Agency for Health and Food Safety (AGES) and as Lecturer of Food Security at the University of Natural Resources and Life Sciences (BOKU). Her work focuses mostly on agricultural long-term experiments and the connection between science and society by engaging stakeholders from the agricultural and school sectors into different kinds of research activities.

Laura Essl studied "Civil Engineering and Water Management" and "Higher Latin American Studies" in Vienna. From 2007 to 2011 she was involved in research projects in Latin America and Asia in the evaluation of alternative water management technologies. Since 2012, she has been working in Austria with different stakeholders in the transfer of information from remote sensing to users and the development of applications for agriculture.

Dr. Francesco Vuolo received the doctoral degree in "Management of Agricultural Resources and Forests" at the University of Naples Federico II, Italy, in 2007. He is currently Senior Scientist at the University of Natural Resources and Life Sciences, Vienna (BOKU), working on satellite-based technologies and their application in agriculture. Since 2003, he has notably been involved in 20 field experimental activities around this topic. He has conducted extensive work and training with farmers in Italy, Spain, France, and Austria. He has spent almost 2 years at the University of Southampton (UK) in 2009–2010, working on the validation of biophysical parameter retrieval algorithms for the Sentinel-2 and -3 missions. He is one of the main developers of BOKU's Sentinel-2 platform running on the Earth Observation Data Centre for Water Resources Monitoring (EODC).

Chapter 5
Precision Weed Management

Sharon A. Clay and J. Anita Dille

Contents

Sharon Clay. My journey in weed science has not been "conventional" in terms of what many weed scientists have done. The terms "nozzle head" and "spray and pray" (or squirt and look) do not fit my research program. When I accepted the assistant professor position in research weed science at South Dakota State University, I was told "you can do anything in the weed science area BUT herbicide trials, those will be done by the extension person." Was this a curse or a blessing? Well, herbicide trials can provide the needed monetary stimulus to be able to do "other things." This lucrative funding source was not available. SO do anything else with limited funding and resources. The first thing I did was examine herbicide movement in soil. This was the early 1990s, and water and environmental quality projects were well funded, and I had multiple years of post-doc experience examining herbicide/soil interactions. The amount of atrazine and alachlor moved offsite after application from different soils and the rate of herbicide metabolism in soil were some of my first funded research projects. These projects are kind of peripheral to weed science, although not well defined at the time. My research road then

S. A. Clay (✉)
South Dakota State University, Brookings, SD, USA
e-mail: sharon.clay@sdstate.edu

J. A. Dille
Kansas State University, Manhattan, KS, USA

© Springer Nature Switzerland AG 2021
T. K. Hamrita (ed.), *Women in Precision Agriculture*, Women in Engineering and Science, https://doi.org/10.1007/978-3-030-49244-1_5

85

led to "site-specific farming." This is where my training in plant physiology and biology have meshed with the weed science discipline. In the late 1990s, a "Site-Specific Management Guideline" manual was proposed by a coordination committee consisting of three SDSU professors (C.G. Carlson, D.E. Clay, and S.A. Clay) and three Phosphate Potash Institute personnel (later International Plant Nutrition Institute – IPNI) (P. Fixen, S. Murrell, and H. Reetz). This set of guidelines (46 in all) spanned a wide array of site-specific topics from setting up and analyzing on-farm experiments, developing management zones for insects, and soil testing and applying variable rate nitrogen technology (Site-Specific Management Guidelines http://www.ipni.net/ssmg; archival web site now managed by International Fertilizer Assoc). Each guideline was no more than a four-page paper written by leading experts and reviewed for "producer-friendly content" to aid producers in the many aspects of site-specific agriculture. Four of the guidelines were specific to weeds and covered weed scouting (Clay and Johnson, n.d., SSMG 15), weed emergence (Forcella SSMG 20), weed biology (Cardina and Doohan SSMG 25), and using remote sensing to develop weed management zones in soybean (Clay et al. 2004; SSMG 42). In 2011, I edited a book GIS in Agriculture: Invasive Species (CRC Press) that used insects, weeds, pathogens, and invasive animals as targets for control or management. Most chapters provided data sets and step-by-step instructions on how to analyze the data set. Two other books that I helped write or co-edit include *Mathematics and Calculations for Agronomists and Soil Scientists* (Clay et al. 2012a, b) and *Practical Mathematics for Precision Farming* (Clay et al. 2017a, b). These were written to help undergraduate students solve a variety of problems from yield loss due to weeds to nitrogen fertilizer application using blends of products.

J. Anita Dille. My interest in weed science has been focused on understanding "why weeds are where they are" in agricultural fields, the ecological impacts of those weeds on the crop, and exploring multiple integrated means for weed management in those fields. I did my PhD research at the University of Nebraska-Lincoln between 1994 and 1998, and I was in a Weed Ecology research group that was exploring many aspects of spatial and temporal distributions of weeds in fields. It was a great environment to explore that research question, resulting in several published papers focused on generating hypothesis to explain weed distributions in fields (Dieleman et al. 2000a, b; Dieleman and Mortensen 1999). These ideas were the scaffold for aspects of my faculty research and teaching appointment at Kansas State University where I accepted an assistant professor position in Weed Ecology in 1999. I included site-specific weed management research in my program as I identified graduate students with the right interest and set of skills. These skills often included an ability to manage large data sets and analytical techniques, some level of comfort with advanced technological equipment (GPS), and understanding of spatial dimensions of the variability observed in agricultural fields, such as soil characteristics, weed populations, and crop yields.

But, we digress, let's get back to the topic of this chapter, precision weed management.

The definition of "weed" is ambiguous and anthropocentric. "Any plant out of place" and "a plant that is not wanted and in competition with cultivated plants"

(Blatchley 1912) are both definitions of the term weed. These definitions strongly suggest that actions should be taken to remove the unwanted plants. However, an alternative definition of weed is "a plant whose beneficial uses have not yet been discovered" (Emerson 1876) (e.g., "one person's junk is another person's treasure"). This implies that if enough effort is put into understanding the plant, it will have benefit to society and no longer be a weed. If this is true, then "weed control" in some cases may no longer be necessary.

Zimdahl (2011) made an argument that weed science, unlike entomology or plant pathology, missed at least 100 yrs of descriptive observation of "weeds" before modern chemical control methods of these pests were possible. For example, the life cycle of destructive insects and diseases were well described prior to the introduction of modern insecticides and fungicides. However, starting in the 1950s with the introduction of 2,4-D, weed science jumped right into "control" mode, as compared with understanding the organisms that could be controlled. Modern science and "better life through chemistry" gave the world the possibility of killing every unwanted plant and keeping the crop. Glyphosate-resistant crops, introduced in the late 1990s, pushed control of every unwanted plant even further, where a single herbicide could control "all" the weeds without damage to the crop. This method was easy, effective, and almost thoughtless. In fact, we were told that weed scientists were no longer relevant to agronomic sciences, because all that was needed to control weeds was a sprayer and a jug of glyphosate. The problem was that shortly after Roundup-resistant soybean was introduced into the Midwestern US market, a study that reported glyphosate-resistant ryegrass (*Lolium rigidum*) in Australia after several years of continuous use was published (Pratley et al. 1999). This resistant species was observed in an orchard situation, and people disregarded the report as an anomalous occurrence rather than a warning for future concern.

Weed perceptions in agronomic systems are tilted toward the "unwanted" definition rather than any beneficial effects such as providing pollinator habitat, soil cover, food for wildlife, and other environmental services. Because they are unwanted, they must be acted upon and, in almost every case, controlled. But, does every plant that is not crop need to be controlled? The decision to control all but crop can be very resource-intensive and, in the end, not provide any extra benefits such as crop yield, quality, or profitability.

If not all plants can be removed from a cultivated crop, are there any rational criteria for leaving vs controlling a noncrop plant? Are there areas in a field that do not require as much resources as others? That is where precision comes into play and differences between precision weed management vs control may be described. What's the difference between these two? Weed management implies that there is an outlined strategy using a coordinated and integrated set of activities to accomplish the goal of reducing weed interference in cropping systems over many years. Weed control is an attempt to eliminate all weeds from an area and stop them from competing with desired plants immediately. Precision is using either management or control techniques at the sub-field level based on a set of criteria. Is it possible? Precision management for any decision should be approached logically and systematically and broken into manageable steps to be successful (Reetz and Fixen n.d.).

Does it make dollars or sense? What information is needed for precision weed management? What have been barriers to precision adoption?

What is precision management? A very broad definition is managing variability in the field to improve profitability while reducing the impact of agriculture on the environment (Shannon et al. 2018). Precision management aims to optimize inputs, minimize costs, and sustain production yields by understanding and managing the spatial and temporal variability that occurs across a field and years (Kitchen and Clay 2018). So, the first step in a good weed management plan is to understand the field variability (Clay et al. 2018). This may involve the newest gadgets and gizmos by getting overviews of a field OR it may start with basic data and information obtained by observation over cropping seasons. Often for weed management, the information (i.e., what weeds are present) led to the "herbicide shunt," which moved from the information (weeds are present) directly to action (spray the weed with a herbicide) (Johnson and Huggins 1999) (Fig. 5.1). In this study, the authors noted that the information about the weed, weed density, herbicide, or other appropriate action could be better used, by first having knowledge about the weed and weed density and implications about the appropriate course of action (does it cause economic loss in the short term? What will happen if no action is taken? What are the longer-term implications of using a specific herbicide; or what other management could be used?) and, finally, wisdom (e.g., what are the risks involved to the environment, crop, net returns, limitations to the actions taken) to better define the appropriate action for optimal resource management.

I asked a person from India if his father, who has farmed the same acreage for over 25 years, uses precision agriculture. "Oh no, it is much too expensive" was the reply. I then asked if his father knows where the problem spots are, where the high-yielding areas are, and where it is too wet and too dry. "Oh certainly" was the reply. Well, if he has the knowledge about these parameters, can't he use this information to optimize inputs? Isn't that information the cornerstone of what precision agriculture should be? Growers often do not need all the gadgets, although they can be useful, but need to start with what they know and add from there.

What are some of the criteria for selective/precision weed management? Targeting plants in specific areas that result in substantial crop loss certainly would be one of the most important factors (Johnson et al. 1995) for control. Another consideration is to decide which, if any, species do not need to be controlled if their presence does not interfere with yield or crop quality. Another question includes,

Fig. 5.1 Understanding needs of precision agriculture weed management. (Modified from Johnson and Huggins 1999)

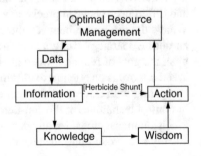

does the crop need to be weed-free for the entire growing season to be productive? If not, when does it make the most sense to strive for weed-free conditions? These and other fundamental questions have been the basis of numerous research projects we have been involved with. Perhaps by understanding more about the biological system of crops and the interactions with different types of stress, the "precision" in "precision weed management and control" becomes more apparent.

5.1 Data

Numerous studies have evaluated multiple aspects of precision weed management for post-emergent herbicide application or mechanical removal.

When I started precision weed research, our team walked 160 acre (65 ha) fields with GPS points taken every 100 ft. (30.4 m), weed quadrats placed every 50 ft. (15 m) on the grid line, and weeds enumerated by species and density. On a 160 A field, this was a 26 by 26 grid, with 6.5 miles (10.5 km) logged and 676 points examined. Then, data were subjected to spatial statistics (kriging) and extrapolated data entered into mapping programs (Clay et al. 2004; Reitsma and Clay 2011). These maps were subjected to further ground truthing for accuracy. Unfortunately, those recording the data already knew that the maps could be inaccurate, as weed patches off the gridlines were evident. Dille et al. (2002) used different spatial interpolators to examine the accuracy of different methods, including inverse distance weighting, point kriging, minimum surface curvature, and multiquadric radial basis function, to develop a weed map. They reported that grid sample spacing and quadrat size were more important than the type of interpolation used, i.e., grids spaced close together and larger quadrats provided more accurate maps. The labor needed to obtain this information at large scales needs to be automated.

In order to get whole field data more quickly, aerial (airplane) imaging using multispectral cameras (blue, green, red, and NIR bands) helped pick out "hot spots," areas where reflectance intensity was unusually high compared to other field areas (Schlemmer et al. n.d.; Dalsted 2011). This helped identify areas that needed "ground-truthed" data. However, the images took about a week to process, and if the pilot was booked for another job, even if conditions were perfect, we did not get an overflight. Satellite imagery also was used; however, pixel size of publicly available images was 30 by 30 m, meaning that a 160 A field may only be covered by 700 pixels, and anomalies may be masked when this large an area is mixed into one pixel color (O'Neill and Dalsted 2011; USDA Forest Service Remote Sensing Applications Center n.d.). More expensive imagery was available that would provide 5 by 5 m or 1 by 1 m pixels (Weis et al. 2008; O'Neill and Dalsted 2011). However, satellite pass times (once every 10 to 14 d), conditions (night fly over, cloudy), and long delays to processing (May or June images delivered the following January) made these remote-sensed images unreliable for real-time use.

Today, drones, and other UAV (UAS) equipment, equipped with cameras and sensors are more commonplace, and their use has been explored to develop site-specific maps of weed patches (Hassanein and El-Sheimy 2018; Haung et al. 2018; Huang et al. 2018a b; Singh et al. 2020). Platform stability, optimizing sensors (resolution, focal length under highly variable environmental illumination conditions) (Romeo et al. 2013), image processing, assessment and prediction, and product delivery have been noted as challenges (Singh et al. 2020). However, data collection and data processing often can be completed within a few hours (Huang et al. 2018a) rather than days to months. This information (if in the correct format) can be uploaded into mapping programs with relative ease. However, even this type of mapping may be too cumbersome for large-scale use. Often these systems are tested on 100 m² areas and are quite accurate (93% or more) in detecting weed patches (Haung et al. 2018a). However, overflights need to have enough overlap of passes (30% or greater) with the images stitched together to get a single image, and 1000 m² may be the target area. Therefore, while a single large field may not be problematic, larger operations that farm 100 s or 1000s of ha may find this method too time-consuming to be of immediate, effective use.

Computer-based imaging techniques for detection of weed infestations across the landscape, using shape, spectral characteristics, and texture, have been reported (Reisinger et al. 2004) with the end goal to automate sprayer technology and/or robotic weeding operations. In some cases, crop vs not crop is the easiest designation. However, using more information, image processing, and classification algorithms provide greater information and discrimination among species.

The information collected about weed species and location, soils, climate, electrical conductivity (EC), and other variables is being gathered by mobile devices, wireless sensor networks, and other devices that do, or will in the near future, connect to data analytic systems, which will be used for decision support. "Traditional" neural network or artificial intelligence will help in data analysis, visualization, and data curation which will create maps (Dille et al. 2011; Barnett et al. 2007; Lopez-Granados 2011) and aid in decision support, helping to define the optimum control method for the weed species and field location (Franco et al. 2017; Haung et al. 2018b).

5.2   Information

Weed species, density, and distribution across locations. The uneven distribution of plants in cultivated fields has been known by producers even when they just began to farm an area. When talking to producers, they often refer to their "problem fields" or "problem areas," and many would have hand-drawn maps showing these areas (Wiles et al. 2007). Patches may increase or decrease depending on climate, management, and other factors but tend to be located in about the same position over 5–10 years (Cardina et al. 1996; Colbach and Forcella 2011; Reitsma and Clay 2011; Clay et al. 1999; 2006). With the introduction of global position system technology (GPS), these hand-drawn maps were formalized putting a site-specific position with latitude and longitude coordinates to weedy areas by species (Dieleman and Mortensen 1999; Cardina et al. 1996; Rocha et al. 2015). Understanding weed biology also is useful when examining distribution. For example, common cockle-bur (*Xanthium strumarium*) and field horsetail (Equesetum arvense) are typically found only in wet areas of the field, whereas common ragweed (*Ambrosia artemisi-ifolia*) is found in the drier shoulder/summit areas of a field.

5.3 Knowledge

Weed emergence patterns. In addition to spatial species distribution maps, the tim-ing of species emergence is a concern (Forcella n.d.; Cardina and Doohan n.d.; Cordeau et al. 2015; Gardarin et al. 2009). Not all "weeds" emerge at the same time. Based on weed biology, there are those that emerge in early spring (C3 species), those that begin emergence in the heat of the summer (C4 species), and some that emerge in fall (winter annuals). But there is a general pattern to the emergence, for example, field pennycress (*Thlaspi arvense*) emerges in early spring or, in some cases, late fall, whereas large crabgrass (*Digitaria sanguinalis*), witchgrass (*Panicum capillare*), and fall panicum (*P. dichotomiflorum*) are first observed in mid- to late June in the northern tier of the USA (Werle et al. 2014a, b). To compli-cate measurements, along with the spatial and temporal variability of weed patches, climate variability, including soil temperatures and water availability, and tillage influence weed emergence and growth (Cordeau et al. 2017, 2018; Davis et al. 2013; Metcalfe et al. 2015; Vleeshouwers and Kropff 2000) and may influence annual species maps. Perennial species, such as Canada thistle (*Cirsium arvense*), at least spatially, may be more stable (Kalivas et al. 2012).

   Why is time of emergence important? First, good weed management programs are based on knowing the species that are present in the field. Knowing the species is critical for management selection, especially if herbicides are used. This will inform the choices for herbicide, rate, and application timing. Thus scouting early may not provide the information on late emerging weeds, whereas if scouted too late, early emerging weeds may be beyond control. Then there are weeds that are of critical importance to control early in their life cycle but may emerge in early to mid-summer, such as Palmer amaranth (*Amaranthus palmeri*) because if this weed gets taller than 4 cm, herbicidal control is poor.

5.4 Wisdom

Expected yield losses by crop/weed species and densities. Crops have a "weed-free period," and if weeds are present during this time, yield losses result (Moriles et al. 2012; Dille 2014; Zimdahl 1988). The weed-free period is influenced by crop, weed species present, and weed density (Zimhahl 1988). However, the timing is always during the early stages of crop development (emergence or early vegetative stage to mid-vegetative stages). Thus, having control that limits weed interference during early crop growth stages is critical (Huiting et al. 2014). This often coincides with weather conditions that are unfavorable for weed management, as soils may be too wet to take equipment over or too windy for herbicide applications. To assuage these problems, pre-emergence herbicides are used that have residual control that will last until post-emergence applications can be safely applied. Site-specific application of residual herbicides using variable rate application technologies may also be appropriate, as the efficacy of soil-applied products is often influenced by soil texture (clay content) and organic matter (Weber et al. 1987; Blumhorst et al. 1990; Rahman and Matthews 1979; Stewart et al. 2010). Higher herbicide rates typically are recommended when clay or organic matter contents are high to achieve the same efficacy as in soils with low clay or organic matter.

5.5 Action

Maps provide images of field problems and help in prescription development. However, as mentioned above, getting a detailed map can be problematic. Sensors on equipment may automate and solve the mapping problem. An array of sensors could be mounted in front of the management equipment, which signal controls to turn on/off the control equipment (Tian et al. 1999). This type of system may be looking at "greenness" and plant shape or just detecting an anomaly that is not present in the crop row (i.e., if it's not crop, then presume it's weed). Machine vision and artificial intelligence needed have been explored (Fernandez-Quintanilla et al. 2018; Partel et al. 2019), but questions remain about how much accuracy is needed in determining the problem (species and density) present (e.g., is it just "weed"? or does species need to be known, should the application be based on presence alone or should density be considered?) (Ali et al. 2013). In addition, when nonselective post-emergence herbicides could be used (e.g., glyphosate, glufosinate), single herbicides could be directly injected into the spray lines to control all species (Qui et al. 1994a, b; Ollila et al. 1990; Goudy et al. 2001). However, weed resistance to both pre- and post-emergence herbicides has become more commonplace (Owen and Zelaya 2005; Heap 2014, 2020; Zimdahl 2010), such that reliance on a single or limited mode-of-action spectrum of herbicides is highly undesirable and solutions other than synthetic herbicides are needed (Bridges 2000; Busi et al. 2013; Green 2014). In the short term, however, with a major reliance on herbicides, the current trend has been to return to multiple selective herbicides with weed species needing to be recognized. Depending on species present, herbicide application may require different booms and injection tanks. When traveling at a normal rate of speed for spraying, there may need to be a lead time to have the system "ready to go" as the machine travels over the area, implying that the sensors should be spaced in front of the nozzles, but how far? This will depend on machine speed and response time of all parts involved (Vondricka 2007; Emmi et al. 2014). Changing to different chemicals "on-the-go" may not be feasible. However, two booms (one for broadleaf control herbicides and the other for grasses) may be used, with one or both spray systems always ready to be activated, although this too may be problematic as some combinations of herbicides are antagonistic to simultaneous applications, with days suggested between applications (Metzenbacher et al. 2015; Zollinger et al. 2014).

Other researchers are moving from strictly herbicide use to more integrated views of weed control using automation. Automated tractors or small robots that would identify weed species using camera sensors and computer technology and then immediately treat the problem area with different types of tools (microtillage, microspray application, clip) for on-the-go control, depending on the plant biology (Ahmad et al. 2014; Young and Giles 2014; Young et al. 2014). Automated weed control using intelligent cultivators for intrarow weed removal in vegetable crops has been successful (Lati et al. 2016; Perez-Ruiz et al. 2014; Fennimore et al. 2016). These automated systems, which use sophisticated technology, although costly today, are becoming more commonplace in high-value vegetable crops (Mosqueda et al. 2017). Automated system use will continue to increase in acceptance as labor and other costs increase (Westwood et al. 2018). Two additional drivers of the change to less herbicide use include (1) increased consumer demand for organically produced food (Shahbandeh 2019) and (2) the increase in herbicide-resistant weed species that will further restrict the use of herbicides and will further accelerate the development of innovative site-specific weed management solutions (Young et al. 2014; Westwood et al. 2018).

5.6 Optimization of resources

When evaluating costs of precision weed management and control and the technology needed, there are numerous factors and forces that are uncertain. When a grower with limited land asks "Where are the best yielding areas?", "Where are the problem areas?", and "What are the problems?", the answers for the best producers were more than likely readily available in hand-drawn maps (Wiles et al. 2007) or just wisdom from multiple years of producing crops on those acres. After one of the first years of mapping and having yield monitor data for a producer's area, we sat down with all the maps (weed maps, EC maps, and yield maps) and discussed the findings with the land-owner at the end of the season. His comment was, "I knew that tile line (for part of the field) was plugged, but I didn't know it was costing me that much yield." The yield map was all he needed to see to have the tile line fixed the very next year.

Technology costs are unknown as well, but likely to get cheaper. When my dad purchased his first calculator in the late 1960s, it could add, subtract, divide, and multiply, all for just over $1000. At the same time, this seemed a bit outrageous, as a good slide rule was about $10. Today, calculators, with more capabilities than the original one my dad bought, are given away and fit in the palm of the hand or smaller, while my dad's took up a good portion of a desk. Today tractors have GPS guidance, autosteer, yield monitors, self-leveling mechanisms, and other gadgets and gizmos that were not commonplace 10 years ago. In 2014 and 2015 surveys

(Clay et al. 2017a, b; Erickson and Widmar 2015), about 58% of producers are using differential corrected global positioning-controlled autosteer to reduce overlaps and skips and improve input placement (Shockley et al. 2011), although using this technology for variable herbicide rate application (Humburg 2003; Rashidi and Mohammadzamani 2011; Grisso et al. 2010) has had less adoption.

Lambert and Lowenberg-DeBoer (2000) summarized over 108 studies that examined the economics of precision farming technology that aided in developing variable rate or precision management decision. For pest (weeds and/or insects) studies, 7 were examined and 6 of these studies reported an economic benefit of using variable rate control depending on weed pressure and patchiness. Returns ranged from $0.01 to $11.67 per acre. In field trials, herbicide usage was decreased by about 25% using variable herbicide rates based on sensor-controlled sprayer compared with broadcast applications (Dammer and Wartenberg 2007) and up to 90% in other studies (Timmermann et al. 2003). Parnell and Bennett (1999) used modeling to examine weed activated spray systems and the feasibility of economic benefit. They concluded that only considering reduced herbicide use to generate profits did not examine the system well enough to determine if benefits were large enough and risks low enough for equipment investment. Over 243 scenarios and different ways to assess savings, the majority of the time broadcast applications resulted in "the best case" as the risk of weed detection errors (i.e., areas should have been sprayed but were not), with subsequent yield loss, was too high to recommend this type of technology (Parnell and Bennett 1999). Maxwell and Luschei (2005) also examined the outcomes of site-specific weed management. They expanded the lens from economic benefit, which they termed as "limited," to ecological benefits. Site-specific weed management lowered the risk of external impact to multiple ecological systems. Food-web, pollinator habitat, water quality, air quality, and other natural systems may benefit from lower pesticide loads, but the economics of these systems is difficult to monetize, so often overlooked. For example, eastern South Dakota typically has 1 million acres of corn annually. Atrazine is applied annually at about 1.5 lb./acre and over 1.5 million lb. of atrazine are used; therefore, even slight reductions, while not having too much impact on the pocketbook, may be helpful for a wide variety of ecosystems. Equipment improvements, microcomputers, and low-cost imagery are providing larger incentives for site-specific weed management although there is still low adoption in most areas.

However, other things are happening that may help precision weed management become the norm. For example, crop improvements will be made to mitigate biotic and abiotic stresses and maintain yield, even in the presence of weeds (Pester et al. 1999; Valliyodan and Nguyen 2006; Jain et al. 2007; Subrahmaniam et al. 2018; Guo et al. 2018). While still in its infancy, the use of transcriptome analysis and gene regulation has changed the understanding of how weed and environmental stresses impact plant growth (Ryan 2000; Moriles et al. 2012; Hansen et al. 2013; Horvath et al. 2007; Guo et al. 2018). For example, before nutrient and water deficiencies can be measured in corn or soybean due to weed presence, these crops have already altered responses in photosynthetic processes and activated genes that may limit future growth (Moriles et al. 2012; Page et al. 2012, 2009; Afifi et al. 2015a; Horvath et al. 2018, 2019). These studies have elucidated genes that, in the future, may be manipulated (either up- or downregulated depending on the overall crop response) to limit the impact of weed stress on plants (Horvath et al. 2015, 2019; Hansen et al. 2013; He et al. 2012; Westwood et al. 2018). Alternatively, understanding the "omics" of stress responses may lead to development of other chemical treatments that alleviate or mitigate stress response (Afifi et al. 2015a, b). Development of crops that exude allelochemicals (through selection, breeding, or genome modification) would provide allelochemical crops (akin to herbicide-resistant crops) a ready defense system against weeds (Westwood et al. 2018) or develop bioherbicides from organisms that already are attacking weeds (such as rusts or other pathogens). These, of course, must be host-specific and not jump to desirable plants.

So what is the future? Weeds will be present and weed control, to maintain sustainable yield, will be needed. I started this work when we manually walked 160 acre fields multiple times per season with crews of 8 or 10 to finish in a day. Then the data were entered into a computer, and Surfer program, with maps generated by the end of the week. Today a single person (or a remote control pilot and a spotter) can fly the same area and have the information within hours. So in the future, you may just sit at your remote sight, push a button, and have all the information downloaded, or you may have a fleet of robots, solar powered, that just search the field looking for unwanted plants and come to a charging station every so often to fully recharge or have swarms of microdrones flying over the field gathering and sending information through cloud technology to larger equipment that is activated. Are these actions possible? If the vision is not articulated, then the possibility does not exist. Will some of the avenues being used today not exist? Absolutely! What ever happened to 8-track tapes, floppy disks, and even CD disks are becoming obsolete. Flash drives that have 256GB capacity can be bought for $30! When computers were developed Mb capacities were often thought to be excessive in storage, who would ever have a need for that much data storage!

In December 2019, a workshop was held on the SDSU campus that focused on workforce development for future agronomic needs with precision agriculture as the goal (author's personal communication). One of the participants who is a service tech provider strongly suggested that all agronomic decisions (what to plant, where and at what time, fertility, weed management, etc.) will be done through regional centers, with every action automated. The workforce will just push buttons to get the selected process started. I disagree and think that there are too many variables to put into algorithms to come up with correct timing and optimization of operations. However, at the same time, there will be continued development of applications that will aid in crop production.

5.7   Unforeseen Consequences (Weed Increases, Labor Decreases, Workforce Education)

Will precision weed management have unforeseen consequences? As with changes in all other areas of endeavor, yes. When draft animals were replaced by tractors, land use changed from pasture to provide feed for the animals, to cultivated acres (Clay et al. 2017a, b). In addition, fossil fuels were needed to power tractors. The harvester changed labor needs from hand picking to more skilled labor for mechanical upkeep. Weed control progressed from hand-hoeing to cultivation to including synthetic herbicides. People moved from small communities to urban areas to earn a living wage due to reductions in the labor force needs. When Roundup Ready technology emerged as the overarching management practice, I was told that "Weed science is dead, what are you going to do now?" One of the unintended consequences of moving to a single herbicide in multiple herbicide-tolerant crops (i.e., glyphosate) has been the evolution of herbicide-resistant weeds (Johnson et al. 2012; Heap and Duke 2018; Heap 2020), and now producers are turning to the older herbicides (typically at increased cost) and tillage (which can increase soil erosion problems and decrease no-till acres). However, herbicide discovery has been stagnant, with no new modes-of-action marketed for over 20 years and few, if any, new herbicides (not just combinations of old chemistries) being developed and marketed in the near term (Das and Mondal 2014; Ruegg et al. 2007). Producers, however, have a real need and keep asking for cheap, environmentally sound management practices. Phytotoxins from plant pathogens and other natural product chemistries are being explored; however, none, to this point, have been both phytotoxic enough and safe enough (based on environmental and human health measurements) to meet the standards for commercial development (Abbas and Duke 2000; Abbas et al. 1998; Duke 1986). No question, researchers in weed science have a lot of work to do!

Producers are faced with uncertainty and risk in five areas (USDA-ERS 2019). The five areas include (1) production risk due to climate and growth of crop can affect both quantity and quality of the commodities; (2) market or price risk, both what is paid for inputs and received for commodity price; (3) financial risk for loan repayment; (4) institutional risk which relates to laws, taxes, and regulations that impact farming; and (5) personal risk such as accidents, personal crises, and other situations that impact a farm business. Clearly, weed management is only one small piece of a very large puzzle. The uncertainty of yield loss due to weed presence varies (Clay et al. 2012a, b). Yield loss due to weeds can range from 0 to 100% depending on weed species, density, time of emergence, crop, soil fertility, and rainfall/irrigation. In addition, if not controlled, weeds in the field this season can add 1000s or 100,000 s of seeds to the soil seedbank (Clay et al. 2005; Norris 1999), which will germinate in future years and, due to dormancy, for unknown lengths of time (Buhler et al. 1997; Forcella et al. 1997). It is not surprising that in most cases, due to the ability to obtain 90% or better control of weeds using herbicides, fields are devoid of any other plants than crop, rather than using an economic or yield loss threshold level for management (Norris 1999). Precision weed management does not guarantee total weed control. The amount of risk (from daredevil to totally risk adverse) that a grower is comfortable with will dictate how much or little precision management will be conducted in their fields.

Another aspect of precision agriculture is understanding and helping students develop the knowledge, skills, and abilities needed to be competent in the next generation of the agricultural workforce (Erickson et al. 2018). Different positions in an agricultural retail business required different skills, which was not unexpected. Areas that were deemed to be of high importance for the precision agricultural specialist included a general knowledge of precision ag technology, effective written and verbal skills, statistics, and operation of precision ag software (e.g., mapping programs, database query, and interfacing programs). "Big data," machine learning, or artificial (or augmented) intelligence is just beginning to provide deeper understanding of many farm-based systems. In 2020, it was estimated that the average farm generated about 250,000 data points per day; however, in 2030 over 2,000,000 data points will be collected, and this value increases to over 4,000,000 in 2036 (Source: OnFarm BI Intelligence Estimates 2015). Clearly, data analytics and data interpretation will be imperative skills. Robotics and driverless machines are being developed, and as the price comes down due to volumes produced, these will need servicing and upkeep.

Based on a survey of retailers (Fausti et al. 2018), the ability to make effective agronomic recommendations was lower for a precision ag specialist but extremely high for the agronomist. It is expected that the agronomist and the precision ag specialist will not work independently, but in tandem, to provide answers to the growers. This survey found that finding qualified applicants to fill most of the positions dealing with precision agriculture were low (in 2015) and there has been a void in higher education programs that needs to be filled (Erickson et al. 2018; Fausti et al. 2018). In 2017, SDSU launched a Precision Ag (PRAG) major and minor degree programs, blending classes to fulfill agronomic knowledge and ag engineering (e.g., mapping, electronics). At this time the impact of these degrees is too new to evaluate, although of the 120 agronomy undergrads about 45% have declared an PRAG minor and there are over 30 PRAG majors. At Kansas State University in Fall 2016, a Precision Agriculture specialization was introduced in the undergraduate Agronomy curriculum. A course in Site-Specific Agriculture had been taught for the past 20 years but is now a key final course of this new curriculum. KSU has successfully graduated and placed students over the past 2 years in excellent precision agriculture careers.

In an editorial piece in 2004, Shaner asked the question is "Precision weed management: the wave of the future or just a passing fad?" (Shaner 2004). He stated that there are numerous problems in both research and economics that needed to be solved and addressed for precision weed management to become THE "conventional" technique. These include collecting data to demonstrate the value of precision weed management and developing cost-effective technologies that can be implemented easily. He concluded by stating that "precision weed management offers too many advantages to the farmer, environment, and society to just become a passing fad."

We are not good at predicting the future. But gleaning information from multiple disciplines, we think that as engineering and biology advance the science and understanding of different processes, the economy of scale will advance precision weed management. There are still further challenges that need to be addressed to improve precision weed management, i.e., control technology, understanding what level of precision is needed for species identification, etc. However, we agree with Dr. Shaner that the numerous advantages of precision weed management from the farmer, environment, and consumer perspectives will push precision to become the future norm.

References

Abbas, H. K., & Duke, S. O. (2000). Phytotoxins from plant pathogens as potential herbicides. *Journal of Toxicology - Toxin Reviews, 14*, 523–543.

Abbas, H. K., Duke, S. O., Merrill, A. H., et al. (1998). Phytotoxicity of australifungin, AAL-toxins and fumonisin B 1to Lemna pausicostata. *Phytochemistry, 47*, 1509–1514.

Afifi, M., Lee, E., Lukens, L., & Swanton, C. (2015a). Maize (*Zea mays*) seeds can detect above-ground weeds: Thiamethoxam alters the view. *Pest Management Science, 71*, 1335–1345.

Afifi, M., Lee, E., Lukens, L., & Swanton, C. (2015b). Thiamethoxam as a seed treatment alters the physiological response of maize (*Zea mays*) seedlings to neighbouring weeds. *Pest Management Science, 71*, 505–524.

Ahmad, M. T., Tang, L., & Steward, B. L. (2014). Automated mechanical weeding. In S. L. Young & F. J. Pierce (Eds.), *Automation: The future of weed control in cropping systems* (pp. 125–138). Dordrecht: Springer.

Ali, A., Streibig, J. C., & Andreasen, C. (2013). Yield loss prediction models based on early estimation of weed pressure. *Crop Protection, 53*, 125–131.

Barnett, D. T., Stohlgren, T. J., Jarnevich, C. S., et al. (2007). The art and science of weed mapping. *Environmental Monitoring and Assessment, 132*, 235–252.

Blatchley, W. S. (1912). *Indian weed book* (p. 191). Indianapolis: Nature Pub. Co.

Blumhorst, M. R., Weber, J. B., & Swain, L. R. (1990). Efficacy of selected herbicides as influenced by soil properties. *Weed Technology, 4*, 279–283.

Bridges, D. C. (2000). Implications of pest-resistant/herbicide tolerant plants for IPM. In G. G. Kennedy & T. B. Sutton (Eds.), *Emerging technologies for integrated pest management: Concepts, research, and implementation* (pp. 141–153). St Paul: APS Press. ISBN:0–89054-246-5.

Buhler, D. D., Hartzler, R. G., & Forcella, F. (1997). Weed seed bank dynamics. *Journal of Crop Production, 1*, 145–168.

Busi, R., Vila-Aiub, M. M., Beckie, H. J., et al. (2013). Herbicide-resistant weeds: From research and knowledge to future needs. *Evolutionary Applications, 6*, 1218–1211.

Cardina, J., & Doohan D. J. (n.d.). SSMG-25. Weed biology and precision farming. In: *Site specific management guidelines* (pp. 4). IPNI http://www.ipni.net/ssmg. Archival copy managed by International Fertilizer Association. Accessed Jan 2020.

Cardina, J., Sparrow, D. H., & McCoy, E. L. (1996). Analysis of spatial distribution of common lambsquarters (*Chenopodium album*) in no-till soybean (*Glycine max*). *Weed Science, 43*, 258–268.

Clay, S. A. (2011). *GIS applications in agriculture. Volume three: Invasive species* (p. 428). Boca Raton: CRC Press.

Clay S. A., & Johnson, G. A. (n.d.). *SSMG-15 Scouting for weeds*. 4 pg. In: Site-Specific management guidelines. IPNI http://www.ipni.net/ssmg. Archival copy managed by International Fertilizer Association. Accessed Jan 2020.

Clay, S. A., Lems, G. J., Clay, D. E., et al. (1999). Sampling weed spatial variability on a fieldwide scale. *Weed Science, 47*, 674–681.

Clay, S. A., Chang, J., Clay, D. E., Reese, C. L., & Dalsted, K. (2004). *SSMG 42 Using remote sensing to develop weed management zones in soybeans*. 4 pg. In: Site-Specific management guidelines. IPNI http://www.ipni.net/ssmg. Archival copy managed by International Fertilizer Association. Accessed Jan 2020.

Clay, S. A., Kleinjan, J., Clay, D. E., et al. (2005). Growth and fecundity of several weed species in corn and soybean. *Agronomy Journal, 97*, 294–302.

Clay, S. A., Kreutner, B., Clay, D. E., et al. (2006). Spatial distribution, temporal stability, and yield loss estimates for annual grasses and common ragweed (*Ambrosia artemisiifolia*) in a corn/soybean production field over nine years. *Weed Science, 54*, 380–390.

Clay, D. E., Carlson, C. G., Clay, S. A., & Murrell, T. S. (Eds.). (2012a). *Mathematics and calculations for agronomists and soil scientists*. Internl Plant Nutrition Instit Norcross GA 238 pp.

Clay, D. E., Carlson, C. G., Clay, S. A., & Murrell, T. S. (Eds.), (2012b). Chapter 21. Using the hyperbolic model as a tool to predict yield losses due to weeds. In: *Mathematics and calculations for agronomists and soil scientists*. Internl Plant Nutrition Instit Norcross GA 238 pp.

Clay, D. E., Clay, S. A., & Bruggeman, S. A. (2017a). *Practical mathematics for precision farming*. ASA/CSSA/SSSA Madison WI 272 pp.

Clay, D. E., Clay, S. A., DeSutter, T., & Reese, C. (2017b). From plows, horses, and harnesses to precision technologies in the North American great plains. In *Oxford Research Encyclopedia of Environment Science*. Oxford: Oxford University Press. https://doi.org/10.1093/acrefore/9780199389414.013.196.

Clay, S. A., French, B. W., & Mathew, F. M. (2018). Pest measurement and management. In D. K. Shannon, D. E. Clay, N. R. Kitchen (Eds.), *Precision agriculture basics* (pp. 93–102). Wiley Press.

Colbach, N., & Forcella, F. F. (2011). Adapting geostatistics to analyze spatial and temporal trends in weed populations. In S. A. Clay (Ed.), *GIS applications in agriculture* (Invasive species) (Vol. 3, pp. 320–371). Boca Raton: CRC-Press.

Cordeau, S., Guillemin, J. P., Reibel, C., & Chauvel, B. (2015). Weed species differ in their ability to emerge in no-till systems that include cover crops. *The Annals of Applied Biology, 3*, 444–455.

Cordeau, S., Smith, R. G., Gallandt, E. R., et al. (2017). Timing of tillage as a driver of weed communities. *Weed Science, 65*, 504–514.

Cordeau, S., Wayman, S., Reibel, C., et al. (2018). Effects of drought on weed emergence and growth vary with the seed burial depth and presence of a cover crop. *Weed Biology and Management, 18*, 12–25.

Dalsted, K. (2011). Introduction: Remote sensing and GIS techniques for the detections, surveillance, and management of invasive species. In S. A. Clay (Ed.), *GIS applications in agriculture* (Invasive species) (Vol. 3, pp. 1–8). Boca Raton: CRC-Press.

Dammer, K.-H., & Wartenberg, G. (2007). Sensor-based weed detection and application of variable herbicide rates in real time. *Crop Protection, 26*, 270–277.

Das, S. K., & Mondal, T. (2014). Mode of action of herbicides and recent trends in development: A reappraisal. *International Inv Journal Agricultural and Soil Science, 2*, 27–32.

Davis, A. S., Clay, S. A., Cardina, J., et al. (2013). Seed burial physical environment explains departures from regional hydrothermal model of giant ragweed (*Ambrosia trifida*) seedling emergence in U.S. Midwest. *Weed Science, 61*, 415–421.

Dieleman, J. A., & Mortensen, D. A. (1999). Characterizing the spatial pattern of *Abutilon theophrasti* seedling patches. *Weed Research, 39*, 455–468.

Dieleman, J. A., Mortensen, D. A., Buhler, D. D., Cambardella, C. A., & Moorman, T. B. (2000a). Identifying associations among site properties and weed species abundance. I. Multivariate analysis. *Weed Science, 48*, 567–575.

Dieleman, J. A., Mortensen, D. A., Buhler, D. D., & Ferguson, R. B. (2000b). Identifying associations among site properties and weed species abundance. II. Hypothesis generation. *Weed Science, 48*, 576–587.

Dille, J. A. (2014). Plant morphology and the critical period of weed control. In S. L. Young & F. J. Pierce (Eds.), *Automation: The future of weed control in cropping systems* (pp. 51–70). Dordrecht: Springer.

Dille, J. A., Milner, M., Groeteke, J. J., Mortensen, D. A., & Williams, M. M., II. (2002). How good is your weed map? A comparison of spatial interpolators. *Weed Science, 51*, 44–55.

Dille, J. A., Vogel, J. W., Rider, T. W., et al. (2011). Creating and using weed maps for site-specific management. In S. A. Clay (Ed.), *GIS applications in agriculture* (Invasive species) (Vol. 3, pp. 405–418). Boca Raton: CRC-Press.

Duke, S. O. (1986). Microbially produced phytotoxins as herbicides – A perspective. *Review Weed Science, 2*, 15–44.

Emerson, R. W. (1876). Fortune of the republic. In: *Miscellanies*. The complete works of Ralph Waldo Emerson (Vol. XI, pp. 509–544). Boston: Houghton Mifflin.

Emmi, L., Gonzalez-de-Soto, M., Pajares, G., & Gonzalez-de Santos, P. (2014). Integrating sensory/actuation systems in agricultural vehicles. *Sensors, 14*, 4014–4049.

Erickson, B., & Widmar, D. (2015). *Precision agricultural services: Dealership survey results.* West Lafayette: The Center for Food and Agricultural Business/Department of Agricultural Economics and the Department of Agronomy/Perdue University.

Erickson, B., Fausti, S., Clay, D., & Clay, S. (2018). Knowledge, skills, and abilities in the precision agriculture workforce: An industry perspective. *Natural Sciences Education, 47*, 1–11.

Fausti, S., Erickson, B., Clay, S., et al. (2018). Educator survey: Do institutions provide the PA education needed by agribusiness? *Journal of Agribusiness, 36*, 41–63.

Fennimore, S. A., Slaughter, D. C., Siemens, M. C., et al. (2016). Technology for automation of weed control in specialty crops. *Weed Technology, 30*, 823–837. https://doi.org/10.1614/WT-D-16-00070.1.

Fernandez-Quintanilla, C., Pena-Barragan, J. M., Andujar, D., et al. (2018). Is the current state-of-the-art of weed monitoring suitable for site-specific weed management in arable crops? *Weed Research, 58*, 259–272.

Forcella, F. (n.d.). SSMG 20. Estimating the timing of weed emergence. In: *Site specific management guidelines* (p. 4). IPNI http://www.ipni.net/ssmg. Archival copy managed by International Fertilizer Association. Accessed Jan 2020.

Forcella, F., Wilson, R. G., Dekker, J., et al. (1997). Weed seed bank emergence across the Corn Belt. *Weed Science, 45*, 67–76.

Franco, C., Pedersen, S. M., Papaharalampos, H., et al. (2017). The value of precision for image-based decision support in weed management. *Precision Agriculture, 18*, 366–382.

Gardarin, A., Durr, C., & Colbach, N. (2009). Which model species for weed seedbank and emergence studies? A review. *Weed Research, 49*, 117–130.

Goudy, H. J., Bennett, K. A., Brown, R. B., & Tardif, F. J. (2001). Evaluation of site-specific management using a direct-injection sprayer. *Weed Science, 49*, 359–366.

Green, J. M. (2014). Current state of herbicides in herbicide-resistant crops. *Pest Management Science, 70*, 1351–1357.

Grisso, R. D., Alley, M. M., Thomason, W., et al. (2010). Precision farming tools: Variable-rate application. *Virginia Cooperative Extension Publication, 442–505*, 16.

Guo, L., Qui, J., Li, L.-F., et al. (2018). Genomic clues for crop-weed interactions and evolution. *Trends in Plant Science, 23*, 1102–1115.

Hansen, S., Clay, S. A., Clay, D. E., et al. (2013). Landscape features impact on soil available water, corn biomass, and gene expression during the late vegetative stage. *Plant Genome, 6*, 2. https://doi.org/10.3835/plantgenome2012.11.0029.

Hassanein, M., & El-Sheimy, N. (2018). *An efficient weed detection procedure using low-cost UAV imagery system for precision agriculture application* (pp. 181–187). International Arch Photogrammetry, Remote Sensing and Spatial Inform Sci XLII-1, 2018 Symposium paper Oct 2018 Karlsruhe, Germany.

Haung, Y., Reddy, K. N., Fletcher, R. S., & Pennington, D. (2018). UAV low-altitude remote sensing for precision weed management. *Weed Technology, 32*, 2–6.

Haung, J., Deng, J., Lan, Y., et al. (2018a). A fully convolutional network for weed mapping of unmanned aerial vehicle (UAV) imagery. *PLoS One, 13*(4), e0196302. https://doi.org/10.1371/journal.pone.0196302.

Haung, J., Deng, J., Lan, Y., et al. (2018b). Accurate weed mapping and prescription may generation based on fully convolutional networks using UAV imagery. *Sensors, 18*, E3299. https://doi.org/10.3390/s18103299.

He, H., Wang, H., Fang, C., et al. (2012). Barnyard grass stress up regulates the biosynthesis of phenolic compounds in allelopathic rice. *Journal of Plant Physiology, 169*, 1747–1753.

Heap, I. (2014). Herbicide resistant weeds. *Integrated Pest Management Reviews, 3*, 281–301.

Heap, I. (2020). *The international survey of herbicide resistant weeds.* Online. Internet. Available at www.weedscience.org. Accessed 14 Jan 2020.

Heap, I., & Duke, S. O. (2018). Overview of glyphosate-resistant weeds worldwide. *Pest Management Science, 74*, 1040–1049.

Horvath, D. P., Llewellyn, D., & Clay, S. A. (2007). Heterologous hybridization of cotton microarrays with velvetleaf (*Abutilon theophrasti*) reveals physiological responses due to corn competition. *Weed Science, 55*, 546–557. https://doi.org/10.1614/WS-07-008.1.

Horvath, D. P., Hansen, S., Moriles-Miller, J. P., et al. (2015). RNA seq reveals weed-induced PIF 3-like as a candidate target to manipulate weed stress response in soybean. *New Phytologist, 207*, 196–210.

Horvath, D. P., Bruggeman, S., Moriles-Miller, J., et al. (2018). Weed presence altered biotic stress and light signaling in maize even when weeds were removed early in the critical weed-free period. *Plant Direct, 2*(4), e00057. https://doi.org/10.1002/pld3.57.

Horvath, D. P., Clay, S. A., Bruggeman, S. A., et al. (2019). Varying weed densities alter the corn transcriptome, highlighting a core set of weed-induced genes and processes with potential for manipulating weed tolerance. *Plant Genome, 12*, 190035. https://doi.org/10.3835/plantgenome2019.05.0035.

Huiting, J., Rotteveel, T., Spoorenberg, P., & van der Weide, R. (2014, March 11–13). Distribution, significance and control of foxtail, *Setaria* spp. and crabgrass, *Digitaria* spp. in the Netherlands, and the situation within Europe. Julius-Kuhn-Archiv 443 pg 671–681 ref 33 In *Proceedings 26th German conference on weed biology and weed control*, Braunschweig, Germany.

Humburg, D. (2003). Site-specific management guidelines: Variable rate equipment technology for weed control (SSMG-7). In D. Clay et al. (Ed.), *Site specific management guidelines*.

Jain, M., Nijhawan, A., Arora, R., et al. (2007). F-box proteins in rice. Genome-wide analysis, classification, temporal and spatial gene expression during panicle and seed development, and regulation by light and abiotic stress. *Plant Physiology, 143*, 1467–1483. https://doi.org/10.1104/pp.106.091900.

Johnson, G. A., & Huggins, D. R. (1999). Knowledge-based decision support strategies. *Journal of Crop Production, 2*, 225–238.

Johnson, G. A., Mortensen, D. A., & Martin, A. R. (1995). A simulation of herbicide use based on weed spatial distribution. *Weed Research, 35*, 197–205.

Johnson, G. A., Breitenbach, F., Behnken, L., et al. (2012). Comparison of herbicide tactics to minimize species shifts and selection pressure in glyphosate-resistant soybean. *Weed Technology, 26*, 189–194.

Kalivas, D. P., Vlachos, C. E., Economou, G., & Dimou, P. (2012). Regional mapping of perennial weeds in cotton with the use of geostatistics. *Weed Science, 60*, 233–243.

Kitchen, N. R., & Clay, S. A. (2018). Understanding and identifying variability. In: D. K. Shannon, D. E. Clay, & N. R. Kitchen (Eds.), *Precision agriculture basics* (pp 13–24). Wiley Press.

Lambert, D., & Lowenberg-DeBoer, J. (2000). *Precision agriculture profitability review*. Site-Specific Management Center, School of Agriculture, Purdue University. www.agriculture.purdue.edu/ssmc/Frames/newsoilsX.pdf. Accessed Jan 2020.

Lati, R. N., Siemens, M. C., Rachuy, J. S., & Fennimore, S. A. (2016). Intrarow weed removal in broccoli and transplanted lettuce with an intelligent cultivator. *Weed Technology, 30*, 655–663. https://doi.org/10.1614/WT-D-15-00179.1.

Lopez-Granados, F. (2011). Weed detection for site-specific weed management: Mapping and real-time approaches. *Weed Research, 51*, 1–11.

Maxwell, B. D., & Luschei, E. C. (2005). Justification for site-specific weed management based on ecology and economics. *Weed Science, 53*, 221–227.

Metcalfe, H., Milne, A. E., Webster, R., et al. (2015). Designing a sampling scheme to reveal correlations between weeds and soil properties at multiple spatial scales. *Weed Research, 56*, 1–15.

Metzenbacher, F. O., Kalsing, A., Dalazen, G., Markus, C., & Merotto, A., Jr. (2015). Antagonism is the predominant effect of herbicide mixtures used for imidazolinone-resistant barnyardgrass (*Echinochloa crus-galli*) control. *Planta Daninha, 33*, 587–597. https://doi.org/10.1590/SO100-83582015000300021.

Moriles, J., Hansen, S., Horvath, D. P., et al. (2012). Microarray and growth analyses identify differences and similarities of early corn response to weeds, shade, and nitrogen stress. *Weed Science, 60*, 158–166.

Mosueda, E., Smith, R., Goorahoo, D., & Shrestha, A. (2017). Automated lettuce thinners reduce labor requirements and increase speed of thinning. *California Agriculture*. https://doi.org/10.1733/ca.2-17a0018.

Norris, R. F. (1999). Ecological implications of using thresholds for weed management. *Journal of Crop Production, 2*, 31–58.

O'Neill, M., & Dalsted, K. (2011). Obtaining spatial data. In S. A. Clay (Ed.), *GIS applications in agriculture* (Invasive species) (Vol. 3, pp. 9–27). Boca Raton: CRC-Press.

Ollila, D. G., Schumacher, J. A., & Frochlich, D. P. (1990). *Integrating field grid sense system with direct injection technology*. ASAE paper no. 90-1628. St. Joseph, Mich: ASAE.

Owen, D. K. K., & Zelaya, A. I. (2005). Herbicide resistant crops and weed resistance to herbicides. *Pest Management Science, 61*, 301–305.

Page, E. R., Tollenaar, M., Lee, E. A., et al. (2009). Does shade avoidance contribute to the critical period for weed control in maize (*Zea mays* L.)? *Weed Research, 49*, 563–571. https://doi.org/10.1111/j.13653180.2009.00735.x.

Page, E. R., Cerrudo, D., Westra, P., et al. (2012). Why early season weed control is important in maize. *Weed Science, 60*, 423–430. https://doi.org/10.1614/WS-D-11-00183.1.

Parnell, D. J., & Bennett, A. L. (1999). Economic feasibility of precision weed management: Is it worth the investment? In R. W. Medd & J. E. Prateley (Eds.), *Precision weed management in crops and pastures, CRC for Weed Management systems Adelaide* (SEA Working Paper 99/02) (pp. 138–148). Adelaide: Agricultural and Resource Economics University of Western Australia.

Partel, V., Kakarla, C., & Ampatzidis, Y. (2019). Development and evaluation of a low-cost and smart technology for precision weed management utilizing artificial intelligence. *Computers and Electronics in Agriculture, 157*, 339–350.

Perez-Ruiz, M., Slaughter, D. C., Fathallah, F. A., et al. (2014). Co-robotic intra-row weed control system. *Biosystems Engineering, 126*, 45–55. http://www.sciencedirect.com/science/article/pii/S1537511014001214.

Pester, T. A., Burnside, O. C., & Orf, J. H. (1999). Increasing crop competitiveness to weeds through crop breeding. *Journal of Crop Production, 2*, 59–76.

Pratley, J., Urwin, N., Stanton, R., et al. (1999). Resistance to glyphosate in *Lolium rigidum*. I. Bioevaluation. *Weed Science, 47*, 405–411.

Qiu, W., Watkins, G. A., Sobolik, C. J., & Shearer, S. A. (1994a). A feasibility study of direct injection variable-rate herbicide application. *Transactions of ASAE, 41*, 291–299.

Qiu, W., Shearer, S. A., & Watkins, G. A. (1994b). *Modeling of variable-rate herbicide application using GIS* (ASAE Paper No 94-3522). St. Joseph: ASAE.

Rahman, A., & Matthews, L. J. (1979). Effect of soil organic matter on the phytotoxicity of thirteen s-triazine herbicides. *Weed Science, 27,* 158–161.

Rashidi, M., & Mohammadzamani, D. (2011). Variable rate herbicide application using GPS and generating a digital management map. In M. Larramendy (Ed.), *Herbicides, theory and applications* (pp. 127–144). Rijeka: InTech Rijeka. ISBN: 978-953-307-975-2.

Reetz, H. F. Jr, & Fixen, P. E. (n.d.). SSMG-28. Strategic approach to site-specific systems. 4 pg. In: *Site-Specific management guidelines.* IPNI http://www.ipni.net/ssmg. Archival copy managed by International Fertilizer Association. Accessed Jan 2020.

Reisinger, P., Lehoczky, E., Nagy, et al. (2004). Database-based precision weed management. *The Journal of Plant Diseases and Protection Sonderheft, XIX,* 467–472.

Reitsma, K., & Clay, S. A. (2011). Using GIS to investigate weed shifts after two cycles of corn/soybean rotation. In S. A. Clay (Ed.), *GIS applications in agriculture* (Invasive species) (Vol. 3, pp. 374–403). Boca Raton: CRC-Press.

Rocha, F. C., Oliveira Neto, A. M., Bottega, E. L., et al. (2015). Weed mapping using techniques of precision agriculture. *Planta daninha, 33,* 157–164. https://doi.org/10.1590/S0100-83582015000100018. Accessed Dec 2019.

Romeo, J., Guerrero, J. M., Montalvo, M., et al. (2013). Camera sensor arrangement for crop/weed detection accuracy in agronomic images. *Sensors, 13,* 4348–4366.

Ruegg, W. T., Quadranti, M., & Zoschke, A. (2007). Herbicide research and development: Challenges and opportunities. *Weed Research, 47,* 271–275.

Ryan, C. A. (2000). The systemin signaling pathway: Differential activation of plant defensive genes. *Biochimica et Biophysica Acta, 1477,* 112–121. https://doi.org/10.1016/S0167-4838(99)00269-1.

Schlemmer, M., Hatfield, J., & Rundquist, D. (n.d.). SSMG-16. Remote sensing: Photographic vs non-photographic systems. 4 pg. In: Site-specific management guidelines. IPNI http://www.ipni.net/ssmg. Archival copy managed by International Fertilizer Association. Accessed Jan 2020.

Shahbandeh, M. (2019). *Growth of organic food and non-food sales in the U.S. 2008–2018.* https://www.statista.com/statistics/244409/organic-sales-growth-in-the-united-states/. Accessed Jan 2020.

Shaner, D. L. (2004). Precision weed management: The wave of the future or just a passing fad? *Phytoparasitica, 32,* 107–110.

Shannon, D. K., Clay, D. E., & Sudduth, K. A. (2018). An introduction to precision agriculture. In: D. E. ShannonClay & N. R. Kitchen (Eds.), *Precision agriculture basics* (pp. 1–12). Wiley Press.

Shockley, J. M., Dillon, C. R., & Stombaugh, T. S. (2011). Whole farm analysis of the influence of autosteer navigation on net returns, risk, and production practices. *Journal of Agricultural and Applied Economics, 431,* 57–75.

Singh, V., Rana, A., Bishop, M., et al. (2020). Chapter three – Unmanned aircraft systems for precision weed detection and management: Prospects and challenges. *Advances in Agronomy, 159,* 93–134.

Site-Specific Management Guidelines. (n.d.). http://www.ipni.net/ssmg. Archival copy managed by International Fertilizer Association. Accessed Jan 2020.

Stewart, C. L., Nurse, R. N., Hamill, A. S., & Sikkema, P. H. (2010). Environment and soil conditions influence pre-and postemergence herbicide efficacy in soybean. *Weed Technology, 24,* 234–243.

Subrahmaniam, H. J., Libourel, C., Journet, E.-P., et al. (2018). The genetics underlying natural variation of plant-plant interactions, a beloved but forgotten member of the family of biotic interactions. *The Plant Journal, 93,* 747–770.

Tian, L., Reid, J. F., & Hummel, J. (1999). Development of a precision sprayer for site specific weed management. *Transactions of ASAE, 42,* 893–900.

Timmermann, C., Gerhards, R., & Kuhbauch, W. (2003). The economic impact of site-specific weed control. *Precision Agriculture, 4*, 249–260.

USDA Forest Service Remote Sensing Applications Center. (n.d.). *A weed manager's guide to remote sensing and GIS – Mapping and monitoring.* RSAC Internet. http://www.fs.fed.us/eng/rsac. 20 Dec 2019.

USDA-ERS. (2019). *Risk in agriculture.* https://www.ers.usda.gov/topics/farm-practices-management/risk-management/risk-in-agriculture.aspx. Accessed Jan 2020.

Valliyodan, B., & Nguyen, H. T. (2006). Understanding regulatory networks and engineering for enhanced drought tolerance in plants. *Current Opinion in Plant Biology, 9*, 1–7. https://doi.org/10.1016/j.pbi.2006.01.019.

Vleeshouwers, L. M., & Kropff, M. J. (2000). Modelling field emergence patterns in arable weeds. *The New Phytologist, 148*, 445–457.

Vondricka, J. (2007). Study on the response time of direct injection systems for variable rate application of Herbicides. PhD thesis. University of Bonn.

Weber, J. B., Tucker, M. R., & Isaac, R. A. (1987). Making herbicide rate recommendations based on soil tests. *Weed Technology, 1*, 41–45.

Weis, M., Gutjahr, C., Ayala, V. R., et al. (2008). Precision farming for weed management: Techniques. *Gesunde Pflanzen, 60*, 171–181.

Werle, R., Bernards, M. L., Arkebauer, T. J., & Lindquist, J. L. (2014a). Environmental triggers of winter annual weed emergence in the midwestern United States. *Weed Science, 62*, 83–96.

Werle, R., Sandell, L. D., Buhler, D. D., Hartzler, R. G., & Lindquist, J. L. (2014b). Predicting emergence of 23 summer annual weed species. *Weed Science, 62*, 267–279.

Westwood, J. H., Charudattan, R., Duke, S. O., et al. (2018). Weed management in 2050: Perspectives on the future of weed science. *Weed Science, 66*, 275–285.

Wiles, L. J., Bobbitt, R., & Westra, P. (2007). Site-specific weed management in growers' fields: Predictions from hand-drawn maps. In F. Pierce & D. E. Clay (Eds.), (pp. 81–103). Boca Raton: GIS Applications in Agriculture CRC-Press.

Young, S. L., & Giles, D. K. (2014). Targeted and microdose chemical applications. In S. L. Young & F. J. Pierce (Eds.), *Automation: The future of weed control in cropping systems* (pp. 139–149). Dordrecht: Springer.

Young, S. L., Meyer, G. E., & Woldt, W. E. (2014). Future directions for automated weed management in precision agriculture. In S. L. Young & F. J. Pierce (Eds.), *Automation: The future of weed control in cropping systems* (pp. 249–259). Dordrecht: Springer.

Zimdahl, R. L. (1988). The concept and application of the critical weed-free period. In M. A. Altieri & M. Liebman (Eds.), *Weed management in agroecosystems: Ecological approaches* (pp. 145–155). Boca Raton: CRC Press.

Zimdahl, R. L. (2010). Chapter 10 the consequences of weed science's pattern of development. In: *A history of weed science in the United States* (pp. 189–207). Elsevier Inc. https://doi.org/10.1016/C2009-0-63984-2.

Zimdahl, R. L. (2011). *Weed science: A Plea for thought – Revisited springer briefs in agriculture* (pp. 73)

Zollinger, R., Howatt, K., & Jenks, B., et al (2014). Avoiding antagonism between grass and broadleaf control herbicides. In: *ND weed control guide*. North Dakota State University W253.

Dr. Sharon A. Clay is a Distinguished Professor of Weed Science at South Dakota State University, Brookings, SD, where she has research and teaching responsibilities. She received a B.S. in Horticulture from the University of Wisconsin-Madison, an M.S. in Plant Science from the University of Idaho, Moscow, and a Ph.D. in Agronomy from the University of Minnesota, St. Paul. She has conducted weed management studies in cropping systems including soybean, corn, wheat, barley, flax, sunflower and safflower, and rangeland. She has also worked on saline soil restoration and is collaborating with soil microbiologists to examine endophytic relationships on plants growing in these high saline soils to determine if beneficials may be transferred to more desirable plants to provide some salinity tolerance. Other topics of research have included weed physiology, the influence of weeds on crop growth and genomic responses, and site-specific weed management strategies. She has had over 45 graduate students (30 women) and has tried to be an advocate for work/life balance for all. She has actively supported these students and her research work with grants that total over $10 million as PI and close to $50 million as Co-PI. Grants have included state, regional, national, and international awards. She has worked in numerous multidisciplinary teams, which have included other pest management specialists, soil scientists, range ecologists, social scientists, data management specialists, geneticists, plant physiologists, and economists.

Dr. Clay has published over 250 peer-reviewed scientific articles, 3 of which received "Paper of the Year" awards from *Weed Science*. Two of the papers dealt with weed/crop transcriptomics and examined up- and downregulation of genes with and without weeds under field conditions. She has served as editor and author of several books dealing with precision agriculture, site-specific management, and calculations needed for agronomic recommendations. She served as Editor of the 2011 CRC Press book "GIS Applications in Agriculture: Vol. 3: Invasive Species" that presents 17 invited chapters from authors from over 15 countries who put together data sets that dealt with mapping invasive species including feral pigs, insects, diseases, and weeds. Dr. Clay has been an associate Editor of *Weed Science*, and Associate Editor and Technical Editor of *Agronomy Journal*.

Dr. Clay has had numerous leadership roles, including first woman president of the American Society of Agronomy (ASA) in 2013, where she helped initiate the Greenfield scholars program, which pairs students with certified crop advisors for workforce development. She also has served on or chaired over 50 national or regional grant review panels including NSF, USDA-NIFA, and others. She has served as President of the South Dakota Chapters of the honorary societies of Sigma Xi and Gamma Sigma Delta, as well as served on numerous university committees. She was elected ASA Fellow in 2009 based on her activity in ASA, and Distinguished Professor status at SDSU in 2017. In 2017, she also was awarded the Gamma Sigma Delta International Distinguished Achievement in Agriculture, and in 2019 she was awarded the SDSU College of Agriculture, Food and Environmental Sciences Outstanding Researcher. She also received the ASA/CSSA/SSSA Women's Mentoring Award in 2019.

Dr. J. Anita Dille is a Professor in Weed Ecology and Assistant Head for Teaching in the Agronomy Department at Kansas State University. She received her Ph.D. in Agronomy from the University of Nebraska-Lincoln (USA) and her B.Sc. (Agr) and M.S. degrees in Crop Science from the University of Guelph (Canada). Her research focuses on understanding the biology and ecology of important weed species found in Kansas agronomic crops, exploring the impacts of cover crops and integrated weed management strategies on their control, and developing site-specific weed management systems. She has mentored 20 M.S. and 7 Ph.D. students and has published more than 50 peer-reviewed publications. She is responsible for teaching courses in weed science, integrated weed management, and advanced weed ecology. She is currently President-elect for the Weed Science Society of America, an international scientific organization, was Associate Editor for *Weed Science* journal, and Past-president of the North Central Weed Science Society, serving as the first female president in its 70 year history.

Chapter 6
Precision Irrigation: An IoT-Enabled Wireless Sensor Network for Smart Irrigation Systems

Sabrine Khriji, Dhouha El Houssaini, Ines Kammoun, and Olfa Kanoun

Contents

S. Khriji (✉)
Measurement and Sensor Technology, Technische Universität Chemnitz, Chemnitz, Germany

LETI Laboratory, National School of Engineers of Sfax, University of Sfax, Sfax, Tunisia
e-mail: sabrinekhriji@ieee.org

D. El Houssaini
Measurement and Sensor Technology, Technische Universität Chemnitz, Chemnitz, Germany

Centre for Research on Microelectronics and Nanotechnology, Technopark of Sousse, Sousse, Tunisia
e-mail: dhouha.el-houssaini@etit.tu-chemnitz.de

I. Kammoun
LETI Laboratory, National School of Engineers of Sfax, University of Sfax, Sfax, Tunisia
e-mail: ines.kammoun@ieee.org

O. Kanoun
Measurement and Sensor Technology, Technische Universität Chemnitz, Chemnitz, Germany
e-mail: kanoun@ieee.org

© Springer Nature Switzerland AG 2021
T. K. Hamrita (ed.), *Women in Precision Agriculture*, Women in Engineering and Science, https://doi.org/10.1007/978-3-030-49244-1_6

6.1 Introduction

Globally, 70% of the world's total freshwater resources is used for irrigated agriculture (Schlosser et al. 2014). This percentage is expected to increase further as a result of demographic growth and food demand increases. Efficient irrigation systems and water management policies have the potential to maintain farm productivity in a time of limited and costly water resources. To this end, precision irrigation is introduced as a new technology that helps farmers to use water efficiently while improving productivity. Smart water management for precision irrigation is critical to boost crop yields, reduce costs, and promote environmental sustainability. Precision irrigation adopts a strategy that considers the spatial and temporal variations of water stress. This concept is enhanced by using data-enabled technologies to implement smart irrigation systems. These systems continuously monitor the water requirements of soil and crops in real time through a network of wireless sensor nodes while recording and handling the collected data in cloud-based applications. Ultimately, the associated information is used to determine the precise time and quantity of water to be applied. Combining water stress management, data-driven analysis, and irrigation scheduling calls for a highly sophisticated architecture involving data streams and advanced devices. For this reason, the Internet of Things (IoT) plays a prominent role in modern agriculture (Khriji et al. 2019; Kanoun et al. 2018).

The IoT paradigm has led to a digital revolution that has disrupted decades of progress in electronics and computing, with unprecedented levels of low-cost data storage (Stergiou et al. 2018), artificial intelligence (Mehmood et al. 2017), mobile computing (Stergiou et al. 2018), software as a service (SaaS) (Dehury and Sahoo 2016), and cloud computing (Asghari et al. 2019). IoT systems use distributed sensing, processing, and communication tools that interact with each other to control and manage physical phenomena (Reddy n.d.). Moreover, IoT devices not only retrieve data from the environment in which they are deployed and interact with the physical world, but they also use preexisting Internet standards to deliver efficient and high-quality smart services for both data transfer and analysis, applications, and communications.

Wireless sensor networks (WSN), as one of the main elements of IoT, include a number of sensor nodes deployed to gather data from sensor fields and transmit it through a gateway to the Internet in an IoT system. Wireless communication reduces the total cost of the network by approximately 80% compared to hardwired alternatives (Bayne et al. 2017; Khriji et al. 2018a). WSN-based systems for agricultural purposes are expected to be user-friendly, straightforward to design, maintain, and update. A further advantage of the use of sensor nodes is to minimize the amount of work involved, thereby reducing the costs associated with deployment, operation, and maintenance. In other words, WSNs provide real-time monitoring of field evolution and crop conditions that support the precision of agriculture by controlling parameters and yielding cyclical data (Jawad et al. 2017). WSNs are scalable in their use, reliable in difficult situations, provide full control in closed-loop operation, and offer more coverage with a high time and space resolution.

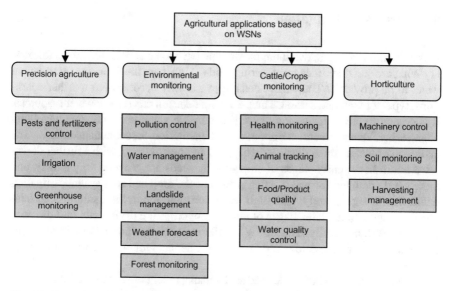

Fig. 6.1 Taxonomy of agricultural applications using WSNs and IoT

A taxonomy of agricultural applications on the basis of WSN is shown in Fig. 6.1. Generally, agricultural applications based on WSN and IoT can be classified into four main classes: precision agriculture, environmental monitoring, cattle/crops monitoring, and horticulture.

According to this taxonomy, notice the importance of water since it is involved in three of the four agricultural application categories. In precision agriculture applications, the control of irrigation is highly recommended to reduce the overuse of water. In some environmental applications, it is interesting to apply different water management policies, which help to determine the upcoming irrigation expectations. Controlling water quality is a critical practice because it directly affects plant nutrition. Sometimes, water cannot properly dissolve fertilizers resulting in a problem in plant growth. In addition, calcium salts in water may produce a white crust of calcium carbonate. Such a layer causes the rapid formation of stones, which can block the irrigation system. For this reason, the focus on this chapter is on irrigation.

The remainder of the paper is organized as follows. Section 6.2 gives an overview of precision irrigation and the key environmental factors affecting crop water needs. Moreover, the most commonly used wireless communication technologies are discussed. Section 6.3 presents an overview of existing relevant smart irrigation system research. Then, in Sect. 6.4, a case study of an IoT-based smart irrigation system is described in detail. Section 6.5 presents important conclusions.

6.2   Precision Irrigation

In the irrigation process, two main factors that have to be considered are as follows:
(1) What are suitable times for watering, and (2) what are the necessary amounts of
water to be provided? The water needs of plants vary according to weather conditions, type of plant, and season. Thus, a wrong decision can reduce water resources.
It is crucial to create technology-based sustainable water use strategies, involving
engineering, management, agronomic, and institution-based enhancements. The
evolution of water demand has demonstrated an upward trend, with a further downward trend. In the early stages of crop growth, seeds are growing very slowly, and
there's little demand for water. After growth of the crop and root system, irrigation
demands rise significantly, with high water consumption in the medium- and late-
fertility stages. In the later stages of cultivation, crops are drying out, and the need
for water is dramatically reduced. When the metabolism and enzymatic reactions
are removed during the cold winter months, the demand for water in the area is relatively low.

Precision irrigation can be defined as the management of water scheduling to suit
the needs of the crops, and it is introduced to adapt to the varying temporal needs of
the plants. The water supplied to the plant is determined by soil, crop, and weather
measurements that reflect the condition of the plant. The main objectives of precision irrigation include increasing water efficiency, reducing energy consumption,
and maximizing crop productivity, using technologies such as wireless sensor networks, network connectivity, mobile devices, real-time monitoring, and remote
sensing.

Figure 6.2 illustrates the common architecture for a precision irrigation system,
including five main layers. The field layer is composed of sensors and actuators. The
sensors are used to measure specific physical parameters including soil moisture,
PH, electrical conductivity, air temperature/humidity, and light intensity.

Furthermore, actuators can be used in such irrigation systems to perform commands from higher tier units. The second layer consists of the wireless communication modules, which can be divided into small-range communication, such as

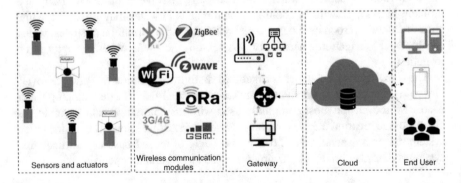

Fig. 6.2 End-to-end communication for precision irrigation

Zigbee and Wi-Fi, and large-range communication such as LoRa and cellular communication. Then, a fog computing layer is introduced to collect sensed data from deployed sensor nodes and execute the required data processing operations for smart aggregation, improving network traffic and minimizing energy costs. This layer is also known as a local gateway, which generates, locally, the fundamental features of the specific process model that are transmitted to the cloud in an efficient and compact form. Thus, the latency and cost of upper layers is significantly reduced. Received data from the gateway unit is continuously transferred to a common cloud computing platform. Therefore, the cloud computing layer provides an opportunity for farmers to use knowledge-based directories to store a wide range of information and insights into farming and equipment options on the market, including all the relevant data. The last layer is the end-user layer or front-end layer, where the results of data analyses are presented to the user. It can provide mobile access and early alerts when an event is triggered.

6.2.1 Environmental Factors in the Field Crops

The productivity and well-being of crops are dependent on several environmental parameters such as air temperature, sunlight, water, gases, and many other environment-related inputs (see Fig. 6.3). Soil moisture is a significant factor in the climatic system that influences water availability to crops. The most common definition of soil moisture is the volumetric water content. Moreover, soil moisture affects the chemical, physical, and biological properties of the soil (Unninayar and Olsen 2008). It is one of the key drivers of the health or stress status of both terrestrial ecosystems and managed systems. Soils are categorized based on their capacity to hold moisture, i.e., minimum moisture, maximum moisture, and optimal moisture. In the appropriate humidity level, crop growth is considered sustainable. In the driest and moistest areas, both enzymatic activities and crop growth have been decreased or interrupted.

Another factor affecting the water demand of field crops is the soil electrical conductivity (EC). EC is a measure that correlates with soil properties such as soil texture, underground characteristics, salinity, and drainage conditions (Ni et al. 2019). Moreover, EC indicates the total level of soluble mineral. A high conductivity refers to a high content of soluble minerals. Reverse osmosis is triggered by an excessively high electrical conductivity of the soil matrix, replacing water in the roots and drying or browning the ends of the roots. Thus, crops are physically and chemically influenced by oversalt in the soil. Soil pH is also an essential factor in the availability of nutrients important for plant growth. It affects the soil concentration and solubility, and the movement of micronutrients in the soil, and therefore the absorption of these substances by crops. A high soil pH can affect plant growth by making some nutrients such as manganese and zinc unavailable.

A low soil pH reduces the availability of phosphorus and other micronutrients (Johnson et al. 2013). Another factor is the soil temperature, which can affect plant

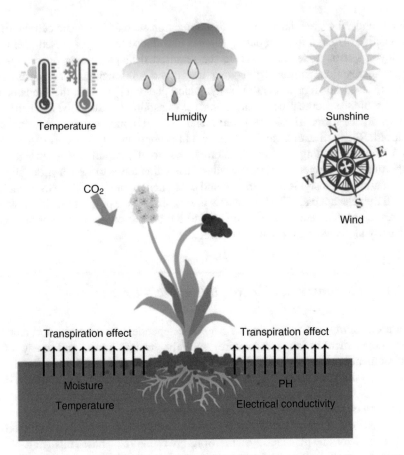

Fig. 6.3 Environmental factors affecting plant productivity

growth by influencing the soil moisture content and plant nutrient availability. Almost all crops are slow growing when the soil temperature is below 90 °C and above 500 °C (Santos et al. 2019).

When the transpiration rate increases, the rate of water absorption by the root also increases. This rate is affected by different factors including air temperature and humidity, light intensity, and wind speed (Mbava et al. 2020). Air temperature has an indirect effect on water absorption by varying leaf transpiration in plants. In fact, the influence of temperature on leaf transpiration increases with the amount of water in the soil. As temperatures increase, enzymatic reactions become more frequent, and the demand for water becomes more significant. As a result, field crops can only satisfy the need for regular nutritional functions. Light intensity is a measure of solar radiation. The more time a field crop is subjected to solar radiation, the warmer the temperature of the crop and soil and the greater the transpiration rate, leading to an additional increase in water demand. Solar radiation boosts the kinetic energy of water in field crops, stimulating water molecules and expanding their uptake. Furthermore, stomata open more widely to permit the presence of more carbon dioxide in the leaf for photosynthesis (Lawson and Vialet-Chabrand 2019). Wind speed induces an increase in airflow, resulting in increased dispersion of water molecules and leaf water transpiration.

The main aim of studying the factors influencing water demand of plants is to identify the most essential sensors that can be used for effective irrigation.

6.2.2 Wireless Communication Technologies

Different communication protocols have been developed in recent decades with the rapid growth of WSN technologies and IoT devices. These protocols have their own specifications based on transmission range, bandwidth, frequency band, data rate, power consumption, cost, and other parameters. In this section, we introduce the most used wireless communication standards for IoT-based agricultural applications. Then a comparison between them is provided in Table 6.1.

Bluetooth Low Energy (BLE) is known as Bluetooth smart technology. And it operates in the 2.4 GHz ISM band. Moreover, BLE is used in some agricultural applications because it is a low-power and low-cost wireless communication technology. It is used to transmit data over short distances up to 10 m and provides data rates up to 200 Kbps. Compared to Bluetooth, BLE can support an unlimited number of nodes in a star topology and has a lower connection time, leading to lower consumed power. However, BLE works only in one-way communication. Thus, BLE can be used in applications where battery lifetime is more critical than high-data transfer speeds. BLE is used in many IoT-based agricultural applications. For instance, in Taskın et al. (2018), the authors designed a new sensor node to monitor ambient light and temperature using the BLE communication standard. In Bjarnason (2017), a moisture and temperature sensor employing BLE is designed to monitor agricultural environments.

Since BLE can support mobile phone accessibility, it is selected to be used as a communication protocol.

Table 6.1 Common wireless communication technologies used in irrigation

Communication protocols	Standard rate	Transmission range	Data rate	Topology	Power
BLE	IEEE 802.15.1	30 m	1–24 mbps	Star, bus	Low
Zigbee	IEEE 802.15.4	10–100 m	20–250 kbps	Star, mesh cluster	Low
Wi-Fi	IEEE 802.11	10–200 m	2–54 mbps	Star	Medium
4G LTE	LTE	<15 km	100 mbps	M2M, star	High
LoRaWAN	IEEE 802.11ah	5–10 km	0.3–50 kbps	M2M, star	Medium

Zigbee is an ultralow power consumption wireless communication technology created by Zigbee Alliance in 2002. It is based on the IEEE standard 802.15.4. This technology supports a small range of communication between 10 and 20 m. Zigbee can be applied for ad hoc, decentralized, and mesh network deployment. A typical Zigbee network includes a coordinator, router, and end devices. The coordinator is responsible for forming the network, storing information, and selecting the used communication channel. The router enlarges the network area coverage by routing traffic between different nodes. The end devices are responsible for data sensing or actuation, and they can transmit or receive data, but they cannot make a route traffic. Zigbee technology is used in many agricultural applications such as automated irrigation and greenhouse. In Chikankar et al. (2015), an automated irrigation system-based Zigbee technology is designed. This system includes two types of nodes, namely, sensor node and actuator. The sensor node is composed of soil moisture, temperature, and air humidity sensor.

Wi-Fi is a wireless local network (WLAN) standard which uses radio wave (RF) in order to enable the exchange of information between two devices and the connection to the Internet. It runs on the IEEE standard 802.11. Wi-Fi uses both the global 2.4 GHz ultrahigh frequency (UHF) and 5 GHz superhigh frequency (SHF) ISM radio bands. This technology provides a communication range of up to 35 m for indoor applications and up to 100 m for outdoor applications, enables the connection of heterogeneous architectures over an ad hoc network, and allows cheaper deployment of LAN networks. Wi-Fi is very common in IoT-based agricultural applications such as greenhouse, smart irrigation, and crop disease monitoring. In Liang et al. (2018), temperature, soil moisture, and intensity sensors are deployed to monitor environmental parameters inside a greenhouse. Sensed data are transmitted to the cloud through Wi-Fi communication technology. In Thakare and Bhagat (2018), a smart irrigation system based on Wi-Fi communication protocol is designed. An Arduino board is connected with soil moisture, soil pH, and temperature sensors. The collected data are then transmitted to the cloud using Wi-Fi technology.

Cellular Communication is mostly used in applications with high-demand data rates. It supports GSM, 2G, 3G, 4G, and 5G cellular communication with high-speed connectivity to the Internet. However, the drawbacks are that it needs high power consumption and requires an expensive infrastructure to be deployed, which makes it not suitable for small and medium networks. The General Packet Radio Service (GPRS) is the data standard for GSM communication and can support Internet (IP) protocol and Point-to-Point Protocol (PPP). It provides high-data transfer speeds up to 171 Kbps. Using packet switching mechanisms, the data are transmitted from many terminals in the network across different channels. GPRS facilitates the usage of Internet applications over mobile networks. In Khelifa et al.

(2015), a smart irrigation system based on the cellular 4G LTE network is designed. Several soil moisture sensors are deployed, building together a ZigBee mesh network. The sensed data are collected on a smart gateway, which forwards it to the farmer via a mobile communication network running 4G LTE.

LoRaWAN is a low-power, long-range network technology. It is a spread spectrum modulation technique derived from Chirp Spread Spectrum (CSS) technology and uses the ISM frequency band. It is mostly used in IoT-based large-scale agricultural applications due to its long-range communication and its low power consumption. For instance, a smart irrigation system is designed in Zhou et al. (2007), where two LoRa boards are deployed. One is connected to soil moisture, temperature, and water-level sensors, while the second board is used to transmit the sensed data to the Internet via Wi-Fi communication protocol.

Implementing the inappropriate network model can result in additional unnecessary infrastructure expenses and failed data acquisition since not all nodes operate properly in all environments. Therefore, the selection of the wireless protocol is a key milestone in the development of a practical and cost-effective IoT solution.

The choice of the right wireless communication protocol enables an IoT application to successfully compete in a booming market. Deciding which protocol to adopt depends largely on the requirements of the application. LoRaWAN can be a greater option for harsh environments, battery-powered network, and large-scale fields with low-data throughput. However, some irrigation systems require higher bandwidth. In this case, the Wi-Fi standard needs to be used with a compromise on power consumption.

6.3   Related Works on Smart Irrigation Systems

Different controlled irrigation techniques, such as sprinkler irrigation and drip irrigation, are recommended to solve the inefficiency of water uses (Barkunan et al. 2019). The quality and quantity of the yield are strongly affected by coping with water shortage as uneven irrigation, or in some cases overflowing water, can reduce soil nutrition and lead to various microbial diseases. It is not easy to estimate precisely the amount of water needed by the crops because many parameters can be involved in making the decision of watering such as type of soil, type of crop, rainfall, and irrigation technique. Existing irrigation systems can be improved by the adoption of new IoT technologies. The use of IoT-based techniques, such as irrigation management using the crop water stress index, can significantly improve crop efficiency (Chaudhry and Garg 2019).

Authors in Sales et al. (2015) proposed a smart irrigation system based on IoT devices to control the use of water in an agricultural field. A cloud farming surveillance system was developed to help the farmer to check soil status. The watering system included mainly three units: a sensor and actuator unit, a cloud unit, and a user application unit. Sensors were responsible for measuring air temperature and soil moisture. The actuator was used to manage the irrigation flow. A gateway node was used to transmit the collected data from sensor nodes to the cloud platform using HTTP protocol.

In Monica et al. (2017), an automated irrigation system was designed including sensing and data storage units. Sensor units were responsible for sensing soil criteria including moisture, temperature, and luminosity. Depending on soil parameters, an automated operation was carried out by switching a motor on and off using threshold values incorporated in the code. The status of the motor is displayed on the home screen by means of GSM messages. The GSM module is interfaced with an Arduino to establish cellular communication between the system and the user.

A mobile-integrated smart irrigation management and monitoring system is presented in Vaishali et al. (2017). The key aim is to control the water supply and monitor plants' parameters including soil moisture, air humidity, and temperature. The irrigation operation depends on a threshold value of soil moisture. The system uses Raspberry Pi, which monitors sensor data and stores them in the cloud. A smartphone is connected to the Raspberry Pi through Bluetooth to turn on and off the motor.

An IoT-based automated agricultural monitoring and control system is implemented in Haque et al. (2019). The system uses a network of several NodeMCUs microcontrollers (ESP8266) to monitor and control multiple systems over the cloud. A Raspberry Pi is used as a gateway node, where a local Message Queuing Telemetry Transport (MQTT) broker is installed to connect all microcontrollers to the cloud. The NodeMCUs constantly monitor the respective states of various elements of the farm and report the data to the central control unit. The user can then take appropriate actions from analyzing this data, i.e., assign their desired tasks to each of the microcontrollers separately by using a mobile-based or Web-based application.

In Gutiérrez et al. (2013), an automated irrigation system is designed. To control the amount of water, threshold values of soil temperature and moisture are stored on the gateway side, which processes sensed data, activates the actuators, and forwards the data to a Web-based application. Solar panels are used to power the entire system. Moreover, the system provided a two-way communication link using a mobile Internet interface, which ensures the data verification and the watering via a website. A wireless sensor unit uses ZigBee technology to send sensed data to the wireless information unit. This latter is equipped with a GPRS module to transfer the data to a Web server.

In Kamienski et al. (2018), a Smart Water Management Platform (SWAMP) employs various IoT-based approaches to design a precision irrigation system in four different countries. Various crop conditions, such as the size of crops and their growth stage, and different weather parameters, such as rainfall and air temperature, are used to estimate the required amount of water based on the location of yield and the type of crops. Moreover, SWAMP is intended to ensure the ability of high-tech devices to work in different circumstances and to be easily reproduced in various places and environments.

A comparative analysis of smart irrigation systems is described in Table 6.2 using various attributes such as the selected hardware, the wireless communication standard, and the programming language. One of the main contributions of this chapter is to design a low-cost, real-time automated irrigation system that helps farmers irrigate their crops in an efficient way.

6.4   Case Study of a Smart Irrigation System

A real-time IoT-based sensor node architecture to monitor soil moisture and temperature is developed to control the quantity of water in agricultural applications (Fig. 6.4). The proposed solution includes five units: sensing, processing, front-end, actuation, and persistence. The sensing unit is composed of a number of wireless nodes integrating soil moisture and temperature sensors. The processing unit consists of a receiver node serially connected to a Raspberry Pi. Sensed data is sent from the receiver to the gateway using the MQTT communication protocol. A python script is written to implement the publisher client to transmit data using the MQTT protocol to specific topics handled by the Mosquitto broker (Pham et al. 2019). A graphical Web interface and a mobile application are designed and implemented to enable the user to monitor the data in real time. Web interface pages are implemented in HTML/CSS and JavaScript, hosted on Raspberry Pi, and served using a lightweight framework called Flask (Aggarwal 2014).

When the moisture levels of the soil reach a certain threshold level, the user is alerted via the mobile or Web interface and can take immediate action to control the motor and reduce the wastage of water and energy. A persistence unit based on the MySQL database is used to store data directly from the publisher and can also be accessed via the Web or mobile application. In the following subsections, we discuss each unit in detail, as well as the implementation of the back-end and front-end functionalities.

6.4.1   Sensing Unit

The sensing unit used in our irrigation control system consists of SEN-13637 resistive-based soil moisture sensors (Sparkfun soil moisture sensor 2019), DS18B20 soil temperature sensors (Ds18b20 datasheet 2019), and a wireless node or mote called panStamp (Khriji et al. 2018b). PanStamp is a small low-power wireless sensor mote programmable from the Arduino IDE. It uses a MSP430 core embedded with a CC1101 RF transceiver, which forms a CC430F5137 System on a Chip (SoC). The unit provides data transfer speeds at a rate up to 600 kbps and can

Table 6.2 Comparison of existing smart irrigation systems

Work	Plant	Hardware		Others	Standards and other technologies	Programming languages
		Sensors	Microcontroller			
Sales et al. (2015)	Peach tree	Soil moisture, weather forecast	MSP430F2274	Electro valve	ZigBee, IEEE 802.15.4, HTTP, GPRS, GPS	JSON, model view controller
Monica et al. (2017)	Groundnut	Moisture, light, temperature	ATMEGA328	Pump	GSM, Bluetooth, Wi-Fi	Sparkfun
Vaishali et al. (2017)	Not mentioned	Temperature, soil moisture	Raspberry Pi	Pump, motor	Bluetooth	Python
Haque et al. (2019)	Not mentioned	Temperature, soil moisture	ESP8266	Raspberry Pi, pump, solar cell	MQTT, Wi-Fi	node.js
Gutiérrez et al. (2013)	Organic sage	Temperature, soil moisture	Pic24FJ64GB004	Photovoltaic panels	GPRS, Zigbee	SQL server, C#

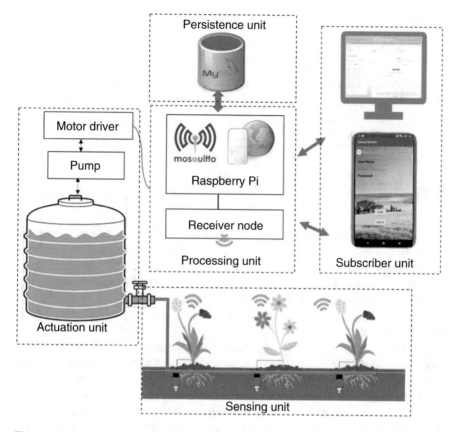

Fig. 6.4 Proposed smart irrigation system architecture

offer a data communication range up to 200 m in outdoor applications. The soil moisture sensor is composed of two pads, together acting as a variable resistor. The more water that is in the soil, the better the conductivity between the pads, resulting in a lower resistance. The temperature sensor has an operating range of −55 C to +125 C. As shown in Fig. 6.5, the panStamp transmitter and the integrated sensors are powered by a battery 5 V DC, and the grounds are commonly connected. The soil moisture sensor signal pin is interfaced with the analog pin of the panStamp.

6.4.2   Processing Unit

In the processing unit, a panStamp receiver node receives sensed data from remote nodes and forwards it to the gateway, which in our case is a Raspberry Pi. This embedded board is used in the proposed prototype due to its common usage in IoT applications and its low cost. The gateway should be fast enough to collect data

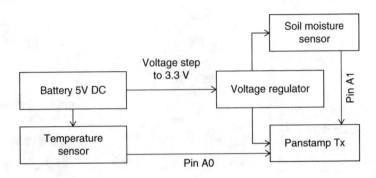

Fig. 6.5 Sensing unit block diagram

from multiple sensor nodes and run algorithms to exchange messages with the cloud. To this end, the choice of the appropriate messaging protocol among the various existing protocols such as MQTT, CoAP, AMQP, and HTTP is highly important. We chose the MQTT protocol as it is distinguished by its short message transmission capability and low bandwidth usage, which makes it suitable for Machine to Machine (M2M) communications of the connected object type (Larmo et al. 2019).

The MQTT network consists of publisher nodes, topics, subscriber nodes, and a broker. Publisher nodes send messages over the topics, which are received by the broker. The broker plays the role of a central unit in every MQTT network and forwards the data to the clients who have subscribed for the topic. MQTT allows devices to concretely send information on a given subject to a server that functions as a message broker. The broker pushes this information to customers who have previously subscribed. An open-source lightweight broker, Mosquitto, is selected which is suitable for low-power single-board computers such as Raspberry Pi. With MQTT, all customer devices that need to communicate with each other must interoperate with the same broker. The latter stores the messages it receives from the sending entities (publishers) and relays them to one or more receivers (subscribers). Messages are sent via a given information channel (or topic). As a result, when the broker receives a published message, it is broadcast to all subscribers, but only those who have subscribed to the given information channel will be able to receive it.

While choosing between the best-suited Web servers for our system, three options were considered: Flask (Aggarwal 2014), Lighttpd (Karayiannis 2019), and Nginx (Siping et al. 2019). Among them, Lighttpd is the most lightweight Web server, which makes it an ideal candidate for Embedded Systems. When compared between Flask and Nginx, both offer a wide range of features like load balancing, fault tolerance, and auto-indexing. However, Flask is the most widely accepted Web server for development purposes and offers better community support. Flask is generally considered a micro-framework that comes with a built-in development server and fast debugger. Since the Web server would not be running any heavy load operations in our case, Flask is the most suitable option for us. Another reason to choose Flask is that it is compatible with other Web servers such as Apache.

The most vital part of making a Web interface is to give the user the ability to interact with the broker in a simplified manner. Any interaction requires a connection from a client with the broker through the Web server. All Web pages in our application are hosted on the Web server, which in our case is our Raspberry Pi. So the Raspberry Pi is serving as a Web server and Mosquitto server. As presented in Fig. 6.6, the idea is that the user can fetch the Web pages from the server using a normal HTTP request method. The Web page can then communicate with the Mosquitto broker using the MQTT protocol. The MQTT client is implemented using Paho MQTT JavaScript library which gives us flexibility to interact with MQTT from the browser using Web sockets.

6.4.3 Subscriber Unit

After the implementation of the core functionalities of the system, the next task was to design and implement a Web interface for the subscriber. Such a Web interface would give the user the ability to either subscribe or publish to topics, monitor data in log form, and visualize data in a plot with historical data in tabular and plot form. The core architecture of the Web interface is based on HTML and CSS, and Python was selected for running operations on the Web server. JavaScript is used to leverage the functionality of Web sockets in Mosquitto. The Eclipse Paho project provides open-source client implementation of MQTT. The libraries are available through source client implementation of MQTT. The libraries are available in JavaScript too, which provides a simpler platform to interface with Mosquitto. Additionally, JavaScript provides better user experience and more control on the browser end. Once the user provides the broker name, port, username, and password, the information is sent to the Mosquitto broker to verify the identity of the subscriber.

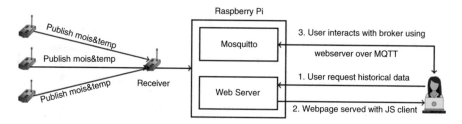

Fig. 6.6 Communication flow in the processing unit

6.4.4   Actuation Unit

In the actuation unit, and as is presented in Fig. 6.7, a 12 V DC water pump integrated with a motor driver is attached to the Raspberry Pi. The 12 V DC motor operates with a current of 200–300 mA and has two silicone tubes used for pumping. One tube is used for fetching water from the tank, and another tube is used for watering the plants at a flow rate of 100 mL/min. A dual H-bridge motor driver, L298N (Yin et al. 2016), is used to control, simultaneously, both speed and direction of the DC motor, which uses a DC power supply of 5 V–46 V with a peak current up to 2A.

The sensed values are transmitted to the server through Raspberry Pi. When the user logs into the mobile application, real-time graphs of soil moisture and temperature sensor data can be viewed. When moisture content in the soil is low, the user can switch ON the motor from the mobile application remotely. This command is received by the server and sets the motor status to "ON." When the pi detects the "ON" status, it sends a signal to the motor driver to turn on the motor and water the plants. The user can turn "OFF" the motor when the soil has enough moisture content. This command is received by the server and sets the motor status to "OFF." When the gateway detects the "OFF" status, it sends a signal to the motor driver to turn off the motor.

6.4.5   Persistence Unit

After transmission of data from wireless nodes to the subscriber node over MQTT, the next milestone is to store data packets in a database for later access. There is a variety of databases, each with its own specifications, strengths, and limits. For our application, we chose to use MySQL for a number of reasons. First, it is a well-known large-scale and open-source relational database that is based around a server/client architecture, which requires a multi-threaded SQL server. Moreover, MySQL supports multiuser features, which along with its scalability makes it a perfect candidate for distributed applications (see Fig. 6.8).

Fig. 6.7 Actuation unit block diagram

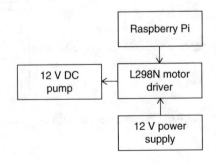

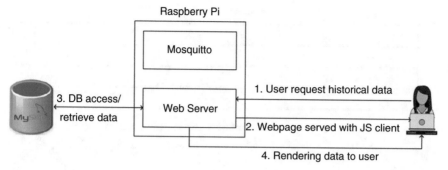

Fig. 6.8 Database access steps

Fig. 6.9 Real-time data monitoring in log form

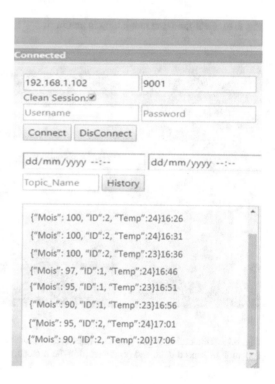

The Web interface enables the user to monitor the historical data of any node in a tabular and plot form. Because we are not concerned about latency here, we can use the HTTP protocol to make a request to the Web server, which in turns makes the query to the database and gives the response back to the user. To fetch data, the user does not need to be connected to the Mosquitto broker because this functionality is directly dealing with the Web server.

Soil moisture and temperature data can be monitored regularly and in real time through the Web application (see Figs. 6.9 and 6.10) so that the user can turn on/off

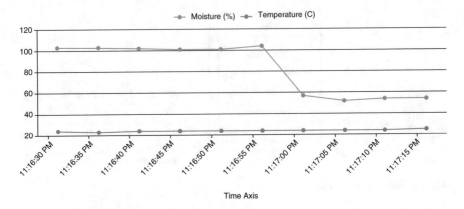

Fig. 6.10 Real-time data monitoring in plot form

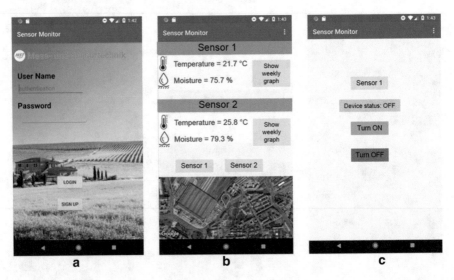

Fig. 6.11 Illustration of the end-user interfaces: (**a**) user authentication layout, (**b**) graphical presentation of sensed data, and (**c**) controlling the motor from the application

the motor in the field, as necessary, through the application from a remote location. The device is updated in the Web server and checked by the Raspberry Pi in real time, and accordingly, it generates a control signal to turn the motor on or off.

Another important feature of the application is the "show weekly graph" button, which directs the user to the weekly graphical google map implemented to check the location of the node. The motor could be turned on or off using the button on an android application (see Fig. 6.11c).

6.5   Conclusion

Water management is a key challenge in agriculture since its availability is a world-wide issue for the forthcoming decades. Therefore, precision irrigation is introduced, as a cutting-edge method, to increase crop productivity and manage the overuse of water with the help of advanced technologies, including wireless sensor networks, cloud computing, and IoT. This chapter highlighted the importance of precision irrigation in agricultural applications. Moreover, the challenging factors affecting the need of crops for water is described to identify the most essential sensors that can be used in such irrigation systems. The most common wireless technologies as well as the general architecture of IoT-based irrigation systems are discussed. The key objective of designing smart irrigation solutions on the basis of IoT and WSN is to save water consumption and reduce manpower, time, and money. For this reason, a cost-effective, real-time IoT-based automated irrigation system is designed using soil temperature and moisture sensors. Sensed data are forwarded to the processing unit (Raspberry Pi), a central node in the system. To collect data from multiple sensor nodes, the lightweight publisher/subscriber messaging transport protocol, MQTT, was used. Mosquitto was used as message broker using the MQTT protocol. When the moisture level of the soil is less than the desired threshold level, the user can take immediate action to give control to the motor through the mobile application, thus reducing the waste of water and energy.

References

Aggarwal, S. (2014). *Flask framework cookbook*. Birmingham: Packt Publishing Ltd.

Asghari, P., Rahmani, A. M., & Javadi, H. H. S. (2019). Internet of things applications: A systematic review. *Computer Networks, 148*, 241–261.

Barkunan, S., Bhanumathi, V., & Sethuram, J. (2019). Smart sensor for automatic drip irrigation system for paddy cultivation. *Computers and Electrical Engineering, 73*, 180–193.

Bayne, K., Damesin, S., & Evans, M. (2017). The internet of things—Wireless sensor networks and their application to forestry. *New Zealand Journal of Forestry, 61*(4), 37–41.

Bjarnason, J. (2017). *Evaluation of Bluetooth low energy in agriculture environments*. Bachelor Thesis, Malmö högskola University.

Chaudhry, S., & Garg, S. (2019). Smart irrigation techniques for water resource management. In *Smart farming Technologies for Sustainable Agricultural Development* (pp. 196–219). Hershey: IGI Global.

Chikankar, P. B., Mehetre, D., & Das, S. (2015). An automatic irrigation system using Zigbee in wireless sensor network. In *2015 International Conference on pervasive computing (ICPC), IEEE* (pp. 1–5).

Dehury, C. K., & Sahoo, P. K. (2016). Design and implementation of a novel service management framework for IoT devices in cloud. *Journal of Systems and Software, 119*, 149–161.

Ds18b20 datasheet., https://datasheets.maximintegrated.com/en/ds/DS18B20.pdf, access (2019).

Gutiérrez, J., Villa-Medina, J. F., Nieto-Garibay, A., & Porta-Gándara, M. Á. (2013). Automated irrigation system using a wireless sensor network and GPRS module. *IEEE Transactions on Instrumentation and Measurement, 63*(1), 166–176.

Haque, M. S. T., Rouf, K. A., Khan, Z. A., Emran, A., & Zishan, M. S. R. (2019). Design and implementation of an IoT based automated agricultural monitoring and control system.

In *2019 International Conference on robotics, electrical and signal processing techniques (ICREST)* (pp. 13–16). Piscataway: IEEE.

Jawad, H. M., Nordin, R., Gharghan, S. K., Jawad, A. M., & Ismail, M. (2017). Energy-efficient wireless sensor networks for precision agriculture: A review. *Sensors, 17*(8), 1781.

Johnson, S. N., Hiltpold, I., & Turlings, T. C. (2013). *Behaviour and physiology of root herbivores* (Vol. 45). Oxford: Elsevier.

Kamienski, C., Soininen, J.-P., Taumberger, M., Fernandes, S., Toscano, A., Cinotti, T. S., Maia, R. F., & Neto, A. T. (2018). Swamp: An IoT-based smart water management platform for precision irrigation in agriculture. In *2018 Global Internet of Things Summit (GIoTS)* (pp. 1–6). Piscataway: IEEE.

Kanoun, O., Keutel, T., Viehweger, C., Zhao, X., Bradai, S., Naifar, S., Trigona, C., Kallel, B., Chaour, I., Bouattour, G., et al. (2018). Next generation wireless energy aware sensors for internet of things: A review. In *2018 15th International Multi-Conference on systems, signals & devices (SSD)* (pp. 1–6). Piscataway: IEEE.

Karayiannis C. (2019) The Lighttpd Web Server. In: Web-Based Projects that Rock the Class. Apress, Berkeley, CA.

Khelifa, B., Amel, D., Amel, B., Mohamed, C., & Tarek, B. (2015). Smart irrigation using internet of things. In *2015 Fourth International Conference on future generation communication technology (FGCT)* (pp. 1–6). Piscataway: IEEE.

Khriji, S., Houssaini, D. E., Kammoun, I., & Kanoun, O. (2018a). Energy-efficient techniques in wireless sensor networks: Technology, components and system design. In *Energy harvesting for wireless sensor networks* (pp. 287–304). Berlin: DE GRUYTER. https://doi.org/10.1515/9783110445053-017.

Khriji, S., Cheour, R., Goetz, M., El Houssaini, D., Kammoun, I., & Kanoun, O. (2018b). Measuring energy consumption of a wireless sensor node during transmission: PanStamp. In *2018 IEEE 32nd International Conference on advanced information networking and applications (AINA)* (pp. 274–280). Piscataway: IEEE.

Khriji, S., El Houssaini, D., Kammoun, I., Besbes, K., & Kanoun, O. (2019). Energy-efficient routing algorithm based on localization and clustering techniques for agricultural applications. *IEEE Aerospace and Electronic Systems Magazine, 34*(3), 56–66.

Larmo, A., Ratilainen, A., & Saarinen, J. (2019). Impact of CoAP and MQTT on NB-IoT system performance. *Sensors, 19*(1), 7.

Lawson, T., & Vialet-Chabrand, S. (2019). Speedy stomata, photosynthesis and plant water use efficiency. *New Phytologist, 221*(1), 93–98.

Liang, M.-H., He, Y.-F., Chen, L.-J., & Du, S.-F. (2018). Greenhouse environment dynamic monitoring system based on WIFI. *IFAC-PapersOnLine, 51*(17), 736–740.

Mbava, N., Mutema, M., Zengeni, R., Shimelis, H., & Chaplot, V. (2020). Factors affecting crop water use efficiency: A worldwide meta-analysis. *Agricultural Water Management, 228*, 105878.

Mehmood, R., Alam, F., Albogami, N. N., Katib, I., Albeshri, A., & Altowaijri, S. M. (2017). Utilearn: A personalised ubiquitous teaching and learning system for smart societies. *IEEE Access, 5*, 2615–2635.

Monica, M., Yeshika, B., Abhishek, G., Sanjay, H., & Dasiga, S. (2017). IoT based control and automation of smart irrigation system: An automated irrigation system using sensors, GSM, Bluetooth and cloud technology. In *2017 International Conference on recent innovations in signal processing and embedded systems (RISE)* (pp. 601–607). Piscataway: IEEE.

Ni, J.-J., Cheng, Y.-F., Bordoloi, S., Bora, H., Wang, Q.-H., Ng, C.-W.-W., & Garg, A. (2019). Investigating plant root effects on soil electrical conductivity: An integrated field monitoring and statistical modelling approach. *Earth Surface Processes and Landforms, 44*(3), 825–839.

Pham, M. L., Nguyen, T. T., & Tran, M. D. (2019). A benchmarking tool for elastic MQTT brokers in IoT applications. *International Journal of Information and Communication Sciences, 4*(4), 70–78.

Reddy, A. S. (n.d.). *Reaping the benefits of the internet of things*. Cognizant Reports, May.

Sales, N., Remédios, O., & Arsenio, A. (2015). Wireless sensor and actuator system for smart irrigation on the cloud. In *2015 IEEE 2nd World Forum on internet of things (WF-IoT)* (pp. 693–698). Piscataway, NJ: IEEE.

Santos, F., Abney, R., Barnes, M., Bogie, N., Ghezzehei, T. A., Jin, L., Moreland, K., Sulman, B. N., & Berhe, A. A. (2019). The role of the physical properties of soil in determining biogeochemical responses to soil warming. In *Ecosystem consequences of soil warming* (pp. 209–244). Elsevier.

Schlosser, C. A., Strzepek, K., Gao, X., Fant, C., Blanc, É., Paltsev, S., Jacoby, H., Reilly, J., & Gueneau, A. (2014). The future of global water stress: An integrated assessment. *Earth's Future, 2*(8), 341–361.

Siping, H., Feng, W., Shejie, L. (2019). Design and optimization of Nginx sever based on LNMP. *DEStech Transactions on Computer Science and Engineering* (iccis).

Sparkfun soil moisture sensor., https://www.sparkfun.com/products/13637, access (2019).

Stergiou, C., Psannis, K. E., Kim, B.-G., & Gupta, B. (2018). Secure integration of IoT and cloud computing. *Future Generation Computer Systems, 78*, 964–975.

Taskın, D., Taskin, C., et al. (2018). Developing a Bluetooth low energy sensor node for greenhouse in precision agriculture as internet of things application. *Advances in Science and Technology Research Journal, 12*, 88–96.

Thakare, S., & Bhagat, P. (2018). Arduino-based smart irrigation using sensors and esp8266 WIFI module. In *2018 Second International Conference on intelligent computing and control systems (ICICCS)* (pp. 1–5). Piscataway: IEEE.

Unninayar, S., & Olsen, L. (2008). *Monitoring, observations, and remote sensing–global dimensions*. In Encyclopedia of Ecology, Jørgensen, SE, Fath BD (eds.). Oxford: Academic Press (pp. 2425 – 2446). [Online]. Available: http://www.sciencedirect.com/science/article/pii/B9780080454054007497

Vaishali, S., Suraj, S., Vignesh, G., Dhivya, S., & Udhayakumar, S. (2017). Mobile integrated smart irrigation management and monitoring system using IoT. In *2017 International Conference on Communication and Signal Processing (ICCSP)* (pp. 2164–2167). Piscataway: IEEE.

Yin, L., Wang, F., Han, S., Li, Y., Sun, H., Lu, Q., Yang, C., & Wang, Q. (2016). Application of drive circuit based on l298n in direct current motor speed control system. In *Advanced laser manufacturing technology* (Vol. 10153, p. 101530N). International Society for Optics and Photonics.

Zhou, Y., Yang, X., Guo, X., Zhou, M., & Wang, L. (2007). A design of greenhouse monitoring & control system based on Zigbee wireless sensor network. In *2007 International Conference on wireless communications, networking and Mobile computing* (pp. 2563–2567). Piscataway: IEEE.

Sabrine Khriji received the diploma of computer science engineering in 2013 and the Embedded Master's degree in 2015 from the national school of engineers of Sfax (ENIS), Tunisia. Since 2016, she has been pursuing the PhD degree in cooperation between the University of Sfax in Tunisia and Technische Universität Chemnitz in Germany (TUC). Her research interests include wireless sensor networks (WSN), Internet of Things (IoT), data aggregation techniques for WSN, low-power embedded systems, information technology, communication and networking, and cloud computing. She has published two book chapters and more than 14 papers in related journals and conferences. She served as a reviewer for different conferences and journals, including the IEEE International Multiconference on Systems, Signals, and Devices (SSD) conference, IEEE WF-IoT, and MDPI applied sciences. She also served on the organizing committee of EnvImeko 2019.

Dhouha El Houssaini received her Diploma in Computer Engineering in 2013 and Master's in Embedded Systems in 2015 from the National School of Engineers of Sfax, Tunisia. Since 2016, she has been pursuing her PhD in Cooperation between the University of Sfax in Tunisia and Technische Universität Chemnitz in Germany. Her research focus is on wireless sensors networks, localization systems and agriculture monitoring, and Internet of Things solutions.

Dr. Inès Kammoun was born in Tunisia in 1975. She received the Engineer Diploma degree from Ecole Polytechnique de Tunis (EPT), Tunisia in 1999 and the Master's and PhD degrees from Telecom ParisTech (National School of Engineering in Telecommunications), Paris, France, in 2000 and 2004, respectively. From 2004 to 2008, she was an assistant professor at Enet'Com (Ex ISECS), Sfax, Tunisia. In 2008, she joined the National Engineering School of Sfax (ENIS), where she is now a full professor. She is a member of the Laboratory of Electronics and Information Technologies (LETI) of ENIS. She is involved in postgraduate research development in wireless communications at ENIS. She also served as TPC co-chair and committee member of several IEEE conferences. She is also an IEEE senior member. Her research interests are in the areas of digital and wireless communications, with special emphasis on MIMO systems, relaying, cooperative networks, cognitive networks, and wireless sensor networks.

Dr. Olfa Kanoun has been a full professor for measurement and sensor technology since 2007 at Technische Universität Chemnitz in Germany. She graduated in electrical engineering at the Technische Universität München in 1996, where she specialized in the field of electronics. Her PhD at the University of the Bundeswehr in Munich was awarded in 2001 by the Commission of Professors in Metrology (AHMT e. V.) in Germany. In 2015, she was awarded by the Tunisian Ministry of Social Affairs for her scientific excellence and outstanding achievements. Her research is focused on measurement methods, sensors, energy harvesting, power aware wireless sensors, and sensor networks. She is a distinguished lecturer of the IEEE Instrumentation and Measurement Society. In 2001, she was the cofounder of the International Multi-conference on Systems, Signals, and Devices (SSD), and in 2008, she initiated the annual International Workshop on Impedance Spectroscopy (IWIS). She is the author or coauthor of 7 books, more than 52 papers in international journals with peer review, 110 papers in proceedings of international conferences, and 6 journal special issues. She is a member of the editorial board of Technisches Messen (De Gruyter) and associate editor of the journal on Digital Signals and Smart Systems (IJDSSS, Inderscience).

Chapter 7
Women Farmer-Breeder Partnerships in Plant Breeding, Seed, and Food Innovations: Experiences from Tigray, Northern Ethiopia

Fetien Abay Abera

Contents

7.1 Introduction

Tigray is the northern regional state of Ethiopia, with 80% of the six million population of the state living in the rural area and largely dependent upon smallholders' agricultural systems for livelihood. On average, nearly 30% of households in the region are female-headed (Meehan 2004). The high occurrence of female-headed households is primarily due to higher numbers of widows having lost husbands either in civil conflicts or naturally due to age gaps and lower remarriage rates of widows than widowers. The female-headed households have constrained livelihoods especially in rural areas due to smaller landholdings, less household labor, and greater difficulty in farming their own land and having access to oxen and labor for

F. A. Abera (✉)
Mekelle University, Mekele, Ethiopia

© Springer Nature Switzerland AG 2021
T. K. Hamrita (ed.), *Women in Precision Agriculture*, Women in Engineering
and Science, https://doi.org/10.1007/978-3-030-49244-1_7

plowing, as well as resources such as fertilizer, improved seed, insecticides, mechanical power, and credit. In addition women are responsible for household chores, child-rearing, and rearing of small livestock, and this keeps them tethered to their households and away from the farm.

Women farmers are more specialized in the selection for certain traits of crops than men. Such gender-based differences are documented in many countries, and the preferences vary with culture and context. Women are skilled in selecting for processing and storage of food grains. They prefer growing crops that are directly consumed without much processing or varieties within crops that have superior taste, are suitable for food preparations, are easily de-huskable, or can be easily preserved and cultivated in poor soils with high weed infestations. For example, in barley varietal selection programs, Tigray women chose varieties that perform well for injera (local flat bread common in Ethiopia) making and can be used immediately without need for winnowing (Abay et al. 2008). Involvement of women farmers helps diversify genotypes both as a risk aversion strategy and as a way of maintaining versatility of the uses of various varieties – thereby strengthening food system resilience. In my research program at Mekelle University, I have always prioritized involving women in barley breeding, seed production, and food processing. This partnership with women collaborators has led to the release of six climate-resilient, highly nutritious (with their beta-glucan, iron, and zinc contents), drought-resistant, early-maturing barley varieties which have reached over 30,000 farmers since 2011, covering over 500 ha. This has resulted in an increase in crop productivity from 2 to 5 ton/hectare. Moreover, the beneficiary farmers, including women, have increased their income by about 30–50%. In this chapter, I will share several examples of the role of women in such activities in the Tigray region of Northern Ethiopia.

The plant breeding team that I'm leading at Mekelle University has multiple types of partnerships with women farmers and women plant breeders. I shall discuss the outcomes of these partnerships in the following three areas, with particular focus on barley improvement: (1) partnership in participatory plant breeding, (2) partnership in the seed sector, and (3) partnership in the food sector and post-harvest handling.

7.1.1 Partnership in Participatory Plant Breeding

Developments in Participatory Approaches in Plant Breeding

Plant breeding is both the art and science of changing the heredity of plants to improve their economic utility. Farmers domesticated plants and selected them for their value, and they were in fact the first plant breeders. They selected plants over generations based on their experience rather than knowledge of genetics. This was called "evolution under the control of man" by N.I. Vavilov. Plant breeding as a science has progressed more recently following molecular developments. However, the identification and selection of desirable types, and letting them differentially interbreed for the next generation, is still dependent on the personal judgment and skill of the breeder. Therefore, combining the experience of farmers and scientific knowledge of plant breeders has advantages in crop improvement and has led to various types of participatory approaches such as participatory plant breeding (PPB) and client-oriented breeding (COB).

There are many advantages of involving farmers in PPB participatory varietal selection (PVS), some of which include:

(a) The selection and testing is in the optimal environment where the variety has to grow ultimately. This reduces genotype x environment interactions (GEI) and failure of the new variety.
(b) Farmers increase the rate of dissemination of preferred types as they save and exchange seeds of the selected varieties.
(c) The yield gap between breeding and extension is reduced because many farmers would have access to the new seed variety before it is formally released and would disseminate it through informal seed systems.

Many authors recognize that the lack of benefit to the resource-poor farmers of marginal areas is due to farmers not being involved in technology development, and they suggested solutions such as the adoption of "farmer-back-farmer" and "farmer-first" approaches (Rhoades and Booth 1982; Chambers and Ghildyal 1985). These concepts led to models of farmer participation in research that were more aimed at empowerment of individuals or communities (Okali et al. 1994). Morris and Bellon (2004) describe five modes of farmer-breeder participation in plant breeding (Fig. 7.1) varying from only farmer breeding to only scientist breeding. Practical participatory breeding methods have used models 2–4. Sperling et al. (2001) defined PPB methods that only used participatory varietal selection (PVS) without involving farmers in the "selection in segregating generations" as in model 4. On the other hand, Witcombe et al. (1996) used model 3 for PPB where farmers make selections in segregating generations as well. However, in the case of model 2, all operations are done on farmers' fields. However, in all practical programs, PVS forms an important component of PPB where farmers test varieties under their own management.

Witcombe et al. (2005) argue that all plant breeding programs are implicitly client-oriented. This way all breeding programs, whether PPB or conventional, fall on a continuum of client orientation rather than discrete classification of PPB and conventional. This is model 3 and is called client-oriented breeding (COB). Witcombe et al. (2005) proposed a parallelism between an industrial innovation model and COB (Fig. 7.2). They described some essential features to make a plant breeding program highly client-oriented which are as follows.

Setting of Goals or Varietal Design

Defining the market for the variety sets the goal of plant breeding for the target market and clients. The participatory rural appraisal (PRA) tool is often used (Chambers, 1997) to define the physical environment (upland, transplanted, irrigated, abiotic and

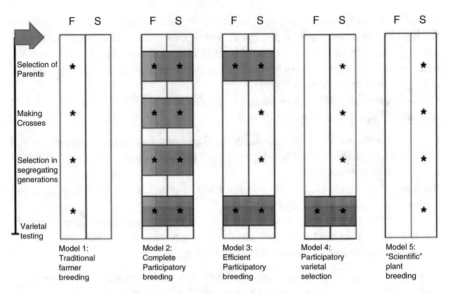

Fig. 7.1 Integrating global and local approaches to plant breeding; F = farmer, S = scientist (Pers. Comm. DS Virk and JR Witcombe). (Adapted from Morris and Bellon 2004)

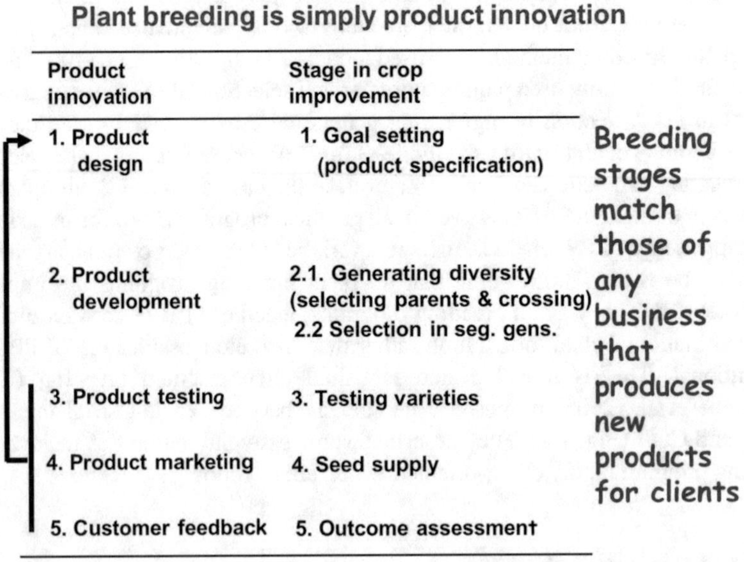

Plant breeding is simply product innovation

Product innovation	Stage in crop improvement	
1. Product design	1. Goal setting (product specification)	Breeding stages match those of any business that produces new products for clients
2. Product development	2.1. Generating diversity (selecting parents & crossing) 2.2 Selection in seg. gens.	
3. Product testing	3. Testing varieties	
4. Product marketing	4. Seed supply	
5. Customer feedback	5. Outcome assessment	

Fig. 7.2 Parallelism between industrial product innovation and plant breeding (Pers. Comm. DS Virk and JR Witcombe)

biotic stresses, use of inputs), socioeconomic environment (category of farmers: poor and smallholder or resource rich), size of market and financial capacity of consumers, target traits in the new variety (color of pericarp in rice (Sthapit et al. 1996)), and ease of threshability (Joshi et al. 2002).

Choosing suitable germplasm: Selecting one locally adapted parent reduces the number of crosses. This is selected from PVS trials and reduces the number of crosses to few and smart or clever crosses to allow growing of large populations of segregating generations. This approach has been shown to be very successful in maize and rice (Witcombe and Virk 2001, Virk et al. 2003, 2005; Witcombe et al. 2003; Gyawali et al. 2002).

Selection in Target Environment

On-station trials are conducted in high-input conditions of package of practice. However, poor farmers hardly apply purchased inputs, and hence the released varieties show high genotype x environment interactions and may fail on farmers' fields. The best strategy will be to match the breeding and testing environments to those of farmers' fields either by simulating them on research stations or by growing the material on farmers' fields and selecting jointly by farmers and breeders in a collaborative participation (Fig. 7.3).

Testing Varieties with Farmers

In the formal system, only a few finally selected varieties by breeders are tested with farmers by the extension workers in a linear extension model. But in the parallel model of farmer participation (Sulaiman and Hall 2002), on-farm testing is conducted simultaneously with research station trials as PVS. An optional feature of COB is making crosses and selection in segregating generations by farmers. Selection in segregating generations by farmers is essential for an empowerment PPB but is optional for collaborative PPB because plant breeders can do this more efficiently. There are many reports favoring selection by farmers in segregating generations (Sperling et al. 1993; Ceccarelli et al. 2001; Gyawali et al. 2002; Joshi et al. 2002), and they indicate that involving farmers can produce better varieties and is a cost-effective way. Consultative selection by farmers along with breeders has also been practiced with advantage (Witcombe et al. 2003; Virk et al. 2003, 2005). Weltzein et al. (2003) and Witcombe et al. (2005) describe circumstances when farmer collaboration is essential.

PVS forms an important component of participatory plant breeding, while COB is important for the selection of parents and testing of advanced lines before release. This way PVS facilitates early supply of varieties while being tested and thus minimizes the delays between formal release and adoption through formal seed chains. PVS also is a means of providing feedback for designing the variety in the PPB/COB programs. Successful plant breeding programs, following principles of COB/PPB,

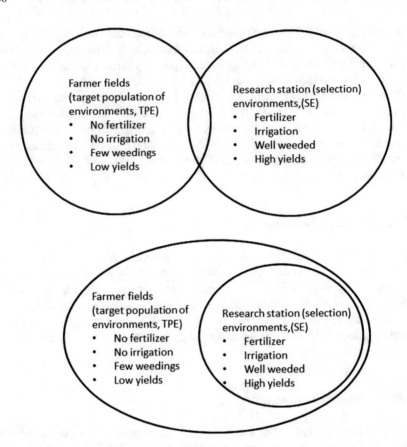

Fig. 7.3 Removing the gap between the farmers' environments (top) and the selection environment of the research stations (bottom) – Witcombe et al. (2005)

were implemented by JR Witcombe and colleagues from Bangor University, UK, in rice in India and Nepal and maize in India. Upland rice breeding by Virk et al. (2003) produced two released varieties from one cross.

Research station trials usually employ the "package of practice" approach designed for obtaining high yield under high-input management. On the other hand, farmers of marginal lands apply lower inputs due to their limited capacity and drought conditions. Varieties for marginal lands will succeed if bred and selected in the optimal environments of farmers' fields or in the nearly simulated conditions on research stations to reduce genotype x environment interactions and increase acceptability of released varieties by farmers.

The most important feature of a successful COB/PPB program is the selection of parents, and making few or clever crosses, and growing large populations. This ensures greater farmer participation as farmers cannot grow many crosses but can grow large populations on their farms. Witcombe et al. (2013) extended the existing models by assuming that the probabilities of success of crosses (P1) decline as more are made and showed fewer crosses were always more efficient than many. In rice breeding in India and Nepal, they obtained a very high success rate in the crosses as measured by the number of varieties released officially or preferred by farmers. In rice breeding in India, a 100% success rate of crosses was achieved by releasing three varieties from three crosses, including one variety developed through marker-assisted selection and farmer participation. Similarly, in rice breeding in Nepal, four varieties were released from six crosses (67% success rate). These are very high success rates compared with one variety from 228 crosses (0.44%) in the national program of Nepal or one variety from 213 crosses by IRRI (0.47%).

7.1.2 Progress in Barley Breeding in Tigray, Northern Ethiopia

Barley (*Hordeum vulgare*) is a staple cereal crop for farmers in Tigray, Northern Ethiopia. This region suffers from recurrent drought and irregular rainfall. Officially recommended "improved" varieties are not suitable for this low-input agricultural system as they are bred in high-input conditions that do not match farmers' field conditions (Fig. 7.3). Therefore, improved varieties have not been accepted by farmers. Farmers' strategy about barley varieties involves avoiding crop failure during drought rather than achieving maximum yield in good seasons. They want early maturity and not only high yield but also good grain quality and the feed value of the straw. Partnership between breeders and resource-poor farmers was thought to be a way to overcome the failures of conventional breeding programs. It is a commonly held belief that farming is men's work, but most practical decisions are made jointly by husband and wife. Rural women in Tigray are generally less educated than men and have limited freedom of movement outside the village. However, women play a key role in barley varietal selection and their preservation following selection in the form of seed. Abay et al. (2008) studied the role of women in decision-making in barley production. Intra-house decisions on the number of varieties to grow, plot allocation, and seed selection were mostly decided jointly (Table 7.1). However, the type of variety to be grown is largely decided by men (72%), but women are typically in charge of seed storage (70%) as well as post-harvest processing (83%).

Table 7.1 Intra-household decision-making (%) on barley production activities

Gender	No. of varieties	Type of variety to grow	Plot allocation for barley	Seed selection	Storage	Post-harvest processing
Women	14	13	16	10	70	83
Men	28	72	37	32	4	4
Both	58	15	47	58	26	13

7.2   Farmer-Breeder Partnership in Varietal Selection of Barley in Tigray

By involving women farmers, we followed PVS approaches (Witcombe et al. 1996; Joshi and Witcombe 1996; Sperling et al. 2001) to provide choices to farmers for the selection of most preferred and locally adapted varieties to the low-input environments of the area. As the first step, we sought varieties that matched farmers' criteria. The selected varieties included 'Himblil' (hulled) and 'Demhay' (hulless) developed by farmers; three local varieties 'Rie', 'Sihumay', and 'Atona'; and four released varieties (MVs), HB-42, Shege, 'Dimtu', and 'Misrach' (released by Holeta and Debre Berhan Agricultural Research Centers), representing both high- and low-input conditions. 'HB42' and 'Shege' (released by Holeta for high-input areas) were deliberately included, being the officially recommended varieties for cultivation in Tigray. On-farm and on-station experiments were conducted in 2005 and 2006 main seasons using the "mother and baby" design (Snapp 1999). These experiments showed that testing of varieties in environments close to farmers' fields is very important for the success of the variety. Correlations between performance on farmers' fields in Tigray and under the more favorable growing conditions in central Ethiopia were poor. Hence, under low-input conditions, the farmer-developed variety (FDV) 'Himblil' was superior to the recommended varieties 'HB-42' and 'Shege'. This indicates a specific adaptation to the prevailing conditions and conforms to the theory that the largest gains for stress are expected from selection in stress conditions (Ceccarelli et al. 1992). The results suggested that Tigray and the central highlands of Ethiopia (where the recommended modern varieties were selected) are different breeding zones (Abay and Bjørnstad 2009). The nature of GEI that cause the cross-overs among genotypes can be positively exploited by selecting different cultivars for the two mega-environments (Ceccarelli et al. 1998; Annicchiarico et al. 2005; Yan et al. 2007). The PVS proved to be a viable method for identifying preferences, constraints, and potentials of varieties in the region. Strong collaborative networks have been established between the farming communities and regional extension systems. This also led to the formation of the "Barley Association" and the sharing of information, ideas, and seeds beyond the trial sites. Farmer-preferred variety 'Himblil' was released through the Ethiopian Variety Release Committee, for multiplication and wider availability to farmers in similar environments. Our research findings can be summarized as follows:

(a) Modern varieties released for high-input conditions were either inferior to or not better than local varieties.
(b) Joint evaluation of varieties with farmers helped in identification and rapid dissemination of new varieties.
(c) Identified varieties can be a source of breeding for specific adaptation.
(d) The need for a participatory and decentralized breeding strategy was confirmed.

In my doctoral research, I investigated the reasons for this adoption failure and introduced the participatory plant breeding (PPB) approach in Tigray to address the situation in marginal environments. The identified genotypes by PVS were used as a source of breeding for specific adaptation and to develop double haploids (Saesa × Himblil: Two-rowed early-escape type, waterlogging-sensitive × six-rowed medium-late, waterlogging-tolerant). This research showed the importance of decentralized breeding that builds on local innovation. The methodology developed through this pioneer work was taken up by other development organizations and projects in Tigray/Ethiopia. The results were widely disseminated and resulted in a follow-up project funded by NUFU from 2007 to 2011 which aimed at using decentralized participatory breeding methods for improving productivity and food security through quality seed in Tigray. The women groups were specifically involved for possibilities of replacing teff in injera with barley and developing fair-trade export of kollo (a traditional roasted barley snack). This project demonstrated new promising barley varieties developed through crosses between 'Himblil' and the major local variety in Tigray, 'Saesa'. The adaptive traits in this cross are being studied further with molecular mapping techniques. This project enhanced collaboration with ICARDA on field testing of a wider array of crosses in Ethiopia and sharing statistical designs that improve experimentation in farmers' fields. We also collaborated on the FP7 EC-funded project "Strategies for Organic and Low-Input Integrated Breeding and Management" (SOLIBAM) with ICARDA and 20 other organizations from 12 European, Mediterranean, and African countries. The experience sharing and international training on PVS and plant breeding has led to the adoption of the strategies in various African countries such as Sudan, Tanzania, Uganda, and others.

7.2.1 Gender-Based Trait Preferences

The differential needs of male and female farmers are reflected in their different preferences for maturation periods, yields, tastes, and colors. Though gender-based differences in preferences are documented in many countries, differences in preference vary with culture and context. Women may well prefer growing crops that are easily prepared for home consumption and require little or no processing or varieties that taste better and can be easily preserved. Because of their time constraints, they may prefer varieties that can be more easily cultivated. Orienting agricultural research to reduce those constraints can make a lasting contribution. For example, where women are labor constrained, affordable mechanization can unleash their productivity. Gender-inclusive research needs to go beyond quantity of production as its only objective, to include taste, food quality, nutrition, processing, resilience, and other characteristics that are particularly important to women. For example, in rural India, Paris et al. (2008) note gender-based differences in preferences for rice varieties whereby women give more importance to traits important for female tasks (such as weed-competitiveness, ease of de-husking and threshing, suitability for food preparation). A study in Rwanda undertaken by the International Center for Tropical

Agriculture (CIAT) demonstrates the importance of recognizing the expertise of female farmers and involving women in participatory plant breeding processes. When 90 Rwandan female farmers evaluated genetic material over a period of 4 growing seasons, the bean varieties selected by the female farmers increased production up to 38% over breeder-selected varieties and outperformed local mixtures 64–89% of the time (Sperling and Berkowitz 1994). Importantly, this study demonstrated the importance of female agricultural knowledge both to researchers and to female farmers themselves. In Tigray, after continuous evaluation and rankings of the different traits of various barley varieties, women chose 'Demhay', a naked barley variety, as the best performer in processing injera and its immediate use without time spent for winnowing (Abay et al. 2008). Their preference for naked over hulled barley is associated with its lower time requirement for processing. It is this recognition of farmers' own innovation and experimentation leading to farmer-developed varieties that has led formal researchers, extension agents, and farmers to work together to design joint experimentation in order to address the challenges of genetic variation and improvement.

The pictures presented in Fig. 7.4 capture the role of women in agricultural research, especially at the grassroots. Women play a lead role in seed selection and undertaking field farm research for varietal and seed selection, improvement, maintenance, and renewal. The photo in Fig. 7.4e captures the innovation of women rice producers in Tselemti Woreda, Tigray. Initially, rice was introduced to the community, and it performed production wise. From the earlier 0.6–0.7 quintals of sorghum per hectare, the newly introduced rice varieties have increased production up to 6–7 tons per hectare. The photo shows that women have a close and microscopic eye in seed size selection during the reproduction state or pre-harvest stages. However, none of the farmers knew how to de-husk it and use the rice. Farmers initially complained "why did they bring this crop to us? We cannot satisfy our hunger by the big heap." Women came up with the innovation of using rice for injera and local beer processing.

Another worth-telling photo story is in Fig. 7.4f, showing a high-quality barley variety nationally released from the PPB trials. This variety has a special character compared to other agronomical good performing PPB lines. It is known for its extra white color and loose spike-own character. This character is especially identified by the women farmer research group involved in the PPB. The main advantage of the variety for the women is that it reduces the labor work cleaning the husks off the grains when cleaning and grading, processing as "kollo" (roasted barley used as a snack), or in a raw form during harvesting. This means reduced time for women processing the grains, hence addresses the gendered dimensions of technology development and dissemination. These innovative women have never gotten any prior training of any kind regarding the processing and use of rice. Yet, they came up with new innovations. This has been a turnaround in the farmers' level of acceptance of rice as one of their major crops. Rice is now regularly produced during the main production season, thanks to women's innovation. It has become the crop of the millennia while adding to the nutrition diversity of local communities. The introduction of rice has also enabled crop rotation with chickpea in waterlogged areas. Currently,

Fig. 7.4 Role of women in agricultural research. Key: **a** and **b** = mother and young ladies in PVS; **c** = family of farmer breeder selecting varieties; **d** = once a daily laborer, she selected one of the PPB lines and multiplied it in a bigger plot on her own; **e** = young lady having a close look while engaging in varietal selection for rice; **f** = high-quality barley variety released in 2012; **g** = a senior woman breeder working with farmers; **h** = young lady grafting fruit seedlings for multiplication and dissemination

farmers are organized into seed producer cooperatives. This also shows the role of women in popularization of new crops and varieties. No crop can survive without utilization, and women demonstrated local innovations for greater utilization. This justifies why women/gender should be part of the larger agricultural growth strategies in any country.

It is clear from our abovementioned studies that farmers require specific adaptation for both edaphic-climatic factors and uses in Tigray. Obviously farmers often selected more than one variety of the same crop.

7.2.2   PVS/PPB Selects for Specific Adaptation

Varieties perform differently in different environments, showing genotype x interactions. Hence, plant breeders select for specific adaptation for obtaining above-average performance in the environment. Varieties with general adaptation perform mediocrely across many environments, but are usually easier to select and breed in the above-average environments. However, poor farmers of drought-prone and marginal areas require varieties that adapt well to their environmental conditions. This is best served by participatory varietal selection. Barley farmers in Tigray have benefitted less from the yield increases of modern breeding because of the lack of specificity in released varieties to their marginal region lands. Abay and Bjørnstad (2009) studied the genotype x environment interactions (GEI) of ten varieties in the Tigray region. The ten varieties included local checks (two farmer-developed varieties, four modern varieties, and three rare local varieties) which were tested over 21 environments. The environments were selected in a participatory mode to sample an adequate number of environments spanning the regional diversity.

Rainfall per month and total nitrogen level were the environmental variables that differentiated all environments into two major groups of environments, the central and northern highlands with less rainfall and poorer soils. In Tigray, rainfall in June and July was negatively correlated with yield, perhaps due to waterlogging problems during the rainy season. The different varieties showed both specific and wide adaptation across environments. Farmer-developed variety 'Himblil' in Tigray was the highest yielding and most stable in the region of its origin. However, it was inferior to improved varieties (Shege and Dimtu) at high yield levels of environments. This clearly showed specific adaptation of Himblil to drought and poor land areas. There was significant association of earliness with grain yield, which showed that there existed a scope for improvement in the existing materials. The results indicated that breeding for drought and waterlogging resistance locations must be practiced by selecting in the target environment. Apparently, involvement of farmers of the region in breeding and selection appeared to be the best strategy to provide stable and high-yielding varieties specifically for the drought-prone and poor areas of Tigray region.

7.3 Impact of Participatory Varietal Selection (PVS)

Impact of PVS was studied in terms of varietal diversification and seed dissemination in three villages (Habes, Mugulat, and Bolenta) in Tigray where a PVS project on barley had been operative (Abay and Halefom 2012). A total of 97 households were randomly selected from 150 participating experimenter farmers. Twenty percent of the respondents were female-headed households. Both qualitative and quantitative data were collected using household survey, matrix ranking, and focused group discussions. The impact on varietal diversity was found because, where farmers grew a few barley varieties before the project in the study area, there was increased varietal richness following introduction of PVS varieties. However, farmers' preference of PVS varieties varied from site to site and there were site-specific adoptions. Out of the PVS varieties, the proportion of Himblil (33%) was the highest, followed by Dimtu (23.4%) and Rie (19.1%). The percentage frequency of other varieties was less than 10% (Fig. 7.5).

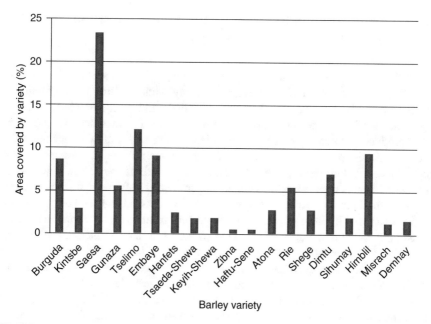

Fig. 7.5 Area share percent allotted to barley varieties PVS participating villages (between 2006 and 2009)

Figure 7.5 shows the richness of the varieties that increased in the study areas. Out of the ten test varieties introduced through the mother and baby trial, eight of them were being grown by farmers in different frequencies.

The analysis of preference matrix ranking showed that Himblil, Misrach, and Dimtu varieties were more preferred by farmers than others due to higher average yield, disease resistance, and number of grains per spike. However, the released variety H42 was the least preferred variety. Collaboration with farmers resulted in climate-resilient, high-yielding, and high nutrition quality in these three barley varieties. Five food barley varieties, Felamit, Hiriti, Adena, Welela, and Fetina, have been released by Mekelle University, and they were developed through participatory plant breeding involving women farmers of Tigray region. These were transgressive segregants from a single cross of two local varieties (Saesa x Himblil). This has resulted in increase in crop productivity from 2 to 5 ton/hectare. Moreover, the beneficiary farmers, including women, have increased their incomes by about 30–50%. Fetina variety is an example of women's role in selecting easy de-husking type of grain, which has a practical advantage as it reduces their work in cleaning, grading, and processing of barley for making "kollo" (roasted barley grain). Time saved from processing the grains is best used in other activities of increasing income and food security.

7.4 Participation in the Seed Sector, Food Innovation, and Post-harvest Loss Management

Mekelle University operated the Tigray Unit of "Integrated Seed Sector Development" (ISSD) project. It operates in four regions (Tigray, Amhara, Oromia, and Southern) of Ethiopia. The aim was to develop a pluralistic and a vibrant seed system in Ethiopia. Seed producer cooperatives were developed. The membership size of the seed producer cooperatives was over 2800 households with 14,000 direct beneficiaries. The producer cooperatives primarily with women farmers have been multiplying and distributing quality seeds to respective Woreda farmers. The objective of ISSD has been to create access to quality superior seed to farmers, thereby improving their food security. Abay and Bjørnstad (2008) reported that seed dissemination of PVS varieties by location was very satisfactory from farmer to farmer and seeds of Himblil and Dimtu reached 47% of experimenter farmers of the study villages in 2009. Seed dissemination in the Mugulat and Habes villages was faster, but none of the experimenter farmers of Bolenta disseminated the seeds any further. This might be due to availability of more varieties adapted to the Bolenta village (Abay and Bjørnstad 2008). The study revealed that Seed Safety through Diversity (SSD) by PVS and Local Seed Business (LSB) activities were encouraging. The "Association of Barley" was founded by a Mugulat local group to promote farmer-based seed dissemination. PVS as part of participatory plant breeding followed by LSB was beneficial for identification and scaling up of seed of preferred varieties in any region. However, the replacement rate of old varieties needs to be appropriately balanced for their genetic conservation in the area.

Women know that for sustainable food production, varietal identity should be maintained. Appearance of more than acceptable variation in the variety indicates the time of its replacement. In sorghum, women in Western Tigray resort to replacement of their sorghum seeds every 3 years or as soon as they observe unacceptable levels of variation in the crop. Women report that appearance of tall and short plants with loose panicles is the right time to replace the seed. Also, women know that appearance of "satan's weed" (zeri seytan) in the sorghum fields is indicative of seed change.

7.4.1 Farmers' Seed/Varietal Choices

The study of local seed systems is important for development. This is more so in Tigray where adoption of modern varieties is low, especially in barley, because they do not perform well under poor management on farmers' fields and lack of irrigation. Farmers thus favor locally adapted varieties which have good buffering capacities for a reasonable performance under adverse conditions. For this reason, we found it important to document and study the local seed system and farmers' role in varietal selection, particularly in the Tigray region, where there is great diversity in local barley varieties. In a study by Abay et al. (2008), farmers' innovations and selection of barley varieties was investigated in the Tigray. A total of seven barley-growing districts and one village from each study district were selected for the study, with the exception of the Gantafeshum District, where two villages were selected because of the greater importance of barley in terms of both coverage and cultural values in this district. At the household level, a total of 240 farmers (30 farmers from each study village) were selected and interviewed. Women household heads and elders were purposely involved to ensure good coverage of diversity in knowledge and seed management. Household surveys sought farmers' views on the values, constraints, and opportunities of growing local varieties of barley. This was supported by focus group and informal discussions with elders, key informants, and women's groups. Case studies were made of local farmers whom the community recognized as barley breeders. Twenty-four barley varieties and their major descriptors were recorded.

Most farmers in Tigray grow barley continuously year after year, except in a few cases where the crop is rotated with legumes. During the season of the survey, 24 varieties were grown. All households grew barley, but no improved variety. About 35% of the farmers were growing one variety, 44% two, 15% three, 6% four, and 1% five varieties, with a mean of 1.93. 26% of farmers that perceived barley production as increasingly associated with its early maturity (50%), multiple uses (35%), and its low requirements for external inputs (15%).

Farmers' decisions to grow local varieties, the land type and area allocated to each, and the other management practices seem to be associated with the role of each local variety in each household's consumption pattern. Additionally the varietal choice is influenced by the household preferences and existing natural resources. Rainfall is the major environmental determinant for yield in dryland farming. At the same time, growing different crops and varieties under varied field conditions helps buffer further against the variation in weather conditions. Some varieties do not grow well on heavy soils, and others do not grow well on lighter sandy soils. Some varieties are better adapted to the lower part of the valley, and others to the higher slopes.

Farmers' Seed Renewal/Selection

In Abay et al. (2008), we found that most surveyed farmers (92%) conserved their barley seeds and practiced mass selection to renew and improve existing varieties. According to our informal discussions, seed selection is based on observations throughout the season, ranging from choosing the right field to condition of standing crops to grain quality characteristics at harvesting. Most farmers gave attention to spike length (89%) and earliness (87%) as reflecting the adaptation to local stresses. Grain color was used as a marker for culinary purposes. After threshing, selection was made among varieties primarily for straw quality (65%), followed by quality for injera (34%) and beverages (33%). According to farmers, selection for disease and pest tolerance (including storability) was done both in the field and during storage. Farmers also selected their seed when multiplying new seed or when recovering seed from drought seasons. Farmers also inspected each plant after harvest. The spikes that were underweight or small-sized were removed from the threshing floor. After threshing, seed for next season was separated by size (large, healthy seeds) at the threshing ground and stored separately in a marked sack or another container.

Farmers in Tigray replaced their seed when its quality was shriveled, diseased, or reduced in size. Barley seed may be saved for 6 years, but the turnover of seed is very high. The variety to be planted as a seed source is allocated to relatively fertile soil and grown in off-season residual moisture. Despite its low yield, this cropping time is recognized as weed-free and produces a bigger grain size compared with main-season production. In this system, Saesa is frequently replaced, and the first seed reproduced in this system is called Saesa to indicate its earliness, and the next three generations are called Wulad (progeny), Salisen (third progeny), and Aregit (old), respectively. These progenies are produced in the main season and are part of the bulk production. After the Aregit generation of seed, the farmers have to seek exchange or purchase of seed from other sources of Saesa or select in that specific season (using the residual moisture).

7.4.2 Women in Post-harvest Loss Management

My involvement as in-country coordinator on USAID's Feed the Future Innovation Lab on Post-Harvest Loss, in collaboration with Kansas State University (USA), has benefitted farmers and five PhD students at both Mekelle and Bahir Dar Universities. Through this project, a national experts team engaged in baseline surveys of post-harvest grain loss in Ethiopia. Our findings showed that women are key players in post-harvest activities, especially storage, in our study sites. As such, women are the first to be informed about storage. The level of participation of women in farming activities before and after harvest is critically important for the productivity of small-scale farms and mitigating against the risks of loss. More than 70% of farm labor is provided by women in Ethiopia. Data from a survey on women's roles in reducing post-harvest losses in four major crops, in four regional states of Ethiopia, shows that during post-harvest activities, women's roles increase up to 88% of labor time, (with men providing 12% of labor), curtailing their contribution to household health, nutrition, and food security. Future policy innovations should, therefore, aim to reduce potential workload of women (Petros et al. 2018). For cash crops like chick-pea and sesame in Ethiopia, post-harvest losses are estimated to average between 10% and 30%, depending on the crop type, but at times they can even be as high as 40%. In some cases, post-harvest losses for sorghum and maize have been especially high, at 15% and 30%, respectively. These losses exacerbate the food insecurity of small-scale farm households. Women are mainly responsible for post-harvest pro-cessing and employ local knowledge that helps reduce post-harvest losses. Women have developed various innovations to reduce post-harvest grain losses (Petros et al. 2018) including:

Mixing seeds of other crops with teff by either simply mixing the seeds together with teff seed and storing them or placing teff at the bottom and top of the storage and putting other seeds in between. Teff is attacked by storage pests less than other grains. Teff is such a fine grain that it limits oxygen movement in the storage, thus reducing the survival chances of storage insect pests, such as weevils.

Smoking storage facilities with chili pepper and blowing the smoke through the storage material or applying pepper powder with the idea that pests will not be comfortable because of the hot property of red peppers.

Placing pumpkin with the crops for its cooling effect as it lowers the in-storage temperature creating less favorable environment for hatching storage pests.

7.4.3 Partnership in the Food Sector: Upgrading Women's Value Chain in Tigray

In order to fight poverty in Ethiopia and to improve food security, one of the key strategies is to promote agricultural product value chain and link farmers to markets to get remunerative prices for their produce. Although there have been similar attempts in the past, unfortunately women's products are not well integrated into these value chain upgrading. An IDRC-funded research project at Mekelle University (PROJECT ID 106956) aims at improving the income of women in remote areas and increase their food security by identifying and implementing strategies that will increase the value chains and market access for local food products developed by women of the Tigray area in Northern Ethiopia. The project focuses on activities of women's cooperatives for improving the income potentials of food products traditionally made by women. With improved market access, it is anticipated that women will be economically empowered and their food security and welfare will be improved. The focus of activities will be to evaluate commercialization potentials of products made from local pulse crops, such as lentils, faba beans, and chickpeas, and to upgrade traditional food product value chains. Lessons learned, when shared with policy makers, will result in policy interventions to ensure increased participation of rural women in remote areas in local and national agricultural markets more effectively and advantageously.

Women in the Tigray region resort to food innovations by preparing food from diverse local and wild sources. These food products are based on mixtures of local cereals and legumes and are the main source of income of households in rural areas and poor urban areas. Thus local varieties of millet sorghum, barley, and legumes are an important source of income for highland farmers of the region. Tigrian women have special techniques of food production and have competitive advantage over other women in Ethiopia. However, total benefits accrued are less than the efforts they apply. These activities have been given little attention by policy makers, and hardly any research and development activities have been undertaken (Abay and Sarah 2009).

The critical research question is therefore how to facilitate women's transition from mere processors of traditional food products to successful active value chain actors? How can we design, test, and promote innovations for upgrading women's traditional food value chains? In view of the above-explained problems, our project is aimed at designing and testing interventions to improve governance of local food product value chains, ensuring that the needs of women food producers and processors are factored into the chain. In addition the project can also help identify strategies for a good linkage (networking) with national customers.

A baseline survey was commissioned by IDRC in Tigray and Amhara regions of Northern Ethiopia under the theme of "upgrading women's food product value chains," in collaboration with the "Institute of Environment, Gender, and Development Studies (IEGDS)" of Mekelle University. Upgrading women's food value chain is transforming trade of products that women specialize in. This requires promoting agricultural value chains by linking farmers in networks and also to the markets. Women's local food value chain provides an entry point so as to integrate women's products into value chain development. These efforts should help in boosting the income, empowerment, and household food security of poor women.

Firstly, a survey targeted the study of challenges to women's products and then outlined innovative strategies that may help in upgrading value chains for local food products and help women better access markets. Secondly, the survey attempted to evaluate opportunities for commercialization and to upgrade traditional food product value chains. The survey covered five districts of Tigray, namely, Hatsebo, Habes, Wukro, Degua Tembien and Hiwane, and one Sokota district from the Amhara region as a comparison site (not in the influence zone of the IDRC project). The project built on some women cooperatives such as Enewani Baltina (Temben) and Selam local food producers' cooperative in Axum that have been established with the financial and facilitation support of the "Women and Food Science Project-MU" (IDRC later followed in organizing them as a group, getting them licensed). The survey considered all women producer groups in the districts as well as informal local food sellers in the local market or outside the local market. Accordingly, 376 women local food producers were included for the baseline study. The main unit of analysis is individual women businesses. The respective staple local food products that were considered for study are Kollo, Tihni, Mitin-Shiro, and Hilbet.

The baseline study has found that there are similarities and differences across products and districts with regard to volume of business, use types, value addition practices, challenges faced, and other issues. The sale of the local foods is the major source of income for most of the women respondents, but it is very low due to many factors. The majority of the local food women producers operate in the informal market, and hence they get little benefit from the support of services by the potential value chain supporters. The local food business is subsistence-based, and traditional value-added product marketing has a low entrepreneurial drive for women businesses. The study revealed that there is poor integration of the value chain players and low attention is given by local governments to the local food market.

Selection and Processing of Barley for Food

Barley in the midlands and highlands of Ethiopia is very important for food security. Food barley is used for varied types of food preparations, and its grain and fodder are important for the animals. We have conducted a number of studies at Mekelle University on the food types and selections in barley, some of which are summarized below.

Varietal Evaluation for Traditional Roasted Barley Foods Kolo and Tihni

Improving the quality of kolo and tihni products through better processing would benefit both producers and consumers. Therefore, in Abraha et al. (2013), we evaluated the impact of processing on the nutritional value of kolo and tihni using eight varieties of barley. We investigated the impact of processing barley grains for kolo and tihni preparation on functional properties and color, varietal differences affecting tihni and kolo processing quality traits, and factors impacting the potential sales of such traditional foods in local, national, and international markets. Results showed significant changes in the chemical and physical properties of raw materials during kolo and tihni processing. Modest variations were found among the eight barley varieties tested for most chemical and physical traits, as well as dehulling properties of hulled barley types. The results confirm that varietal traits such as large grain size and white seed color are important in the selection of raw materials for kolo and tihni processing. Further study on widely divergent genotypes in chemical and physical traits is needed to identify more precisely the breeding goals for quality characteristics of barley used for kolo and tihni processing.

Evaluation of Barley Varieties for Injera

Although tef grain is at least twice as expensive as other cereals, injera from tef is most preferred and consumed daily by the majority of people. This is due to its softer texture, its preferred taste, and especially its color, ranging from purple to "very white," the latter fetching at least 20% higher price that normal "white" types (personal observations). Due to high domestic grain prices, exports of tef from Ethiopia are banned, but some fresh injera made from tef is being exported to the Ethiopian diaspora. Although this positively impacts foreign currency earning, it further increases the price of tef, making tef injera unaffordable for low-income citizens that depend on it as a daily staple. Considering the high price of tef grain in the local market and its low yield potential, searching for a less expensive grain such as barley as a substitute to make injera with comparable quality has become very important.

In the Tigray region in Northern Ethiopia, barley is widely cultivated with an average yield of 1.4 t ha-1. It is the most dependable crop under low-input and semiarid-growing conditions, and farmers emphasize different traits such as grain yield, maturity, and value for injera and other end uses when choosing varieties

(Abay et al. 2008). This has fostered an interest in developing barley varieties with improved injera quality. There are indications that injera quality may vary between cultivars. The recently released variety Himblil, for example, was selected for its superior quality injera. The origin of Himblil is interesting in that the wife of the farmer breeder, Mr. Kahsay Negash, noticed good injera quality in the landrace from which it was extracted (Abay et al. 2008; Abay and Bjørnstad 2009). It is then of interest to validate these observations and to see if injera quality is heritable. Therefore, Saesa, an early-maturing and drought-tolerant cultivar with acceptable injera quality, was crossed with Himblil, a late-maturing, higher-yielding, and high-quality cultivar. There are no previous investigations of genetic variation in barley for injera making quality. Therefore, in Abraha et al. (2013), we conducted a study to (i) compare a set of nine commonly grown barley varieties with Himblil and two known starch mutants (waxy and high amylose) for sensory injera and grain quality, (ii) evaluate 14 F3:7 families from the Saesa × Himblil (S × H) cross, and (iii) analyze injera sensory quality in both sets of genotypes in relation to grain composition and physical grain quality.

Participatory Sensory Evaluation of Injera Quality

A total of 20 panelists (12 men and 8 women, 10 at each site, from 36 to 72 year old) participated in the injera evaluations (Fig. 7.6). The sensory evaluation consisted of scoring one trait at a time, on a scale of 1 to 5 (1 represents very poor, 2 represents poor, 3 represents acceptable, 4 represents very good, and 5 represents excellent). The range within a given sensory attribute was demonstrated and explained before the tests. Individual evaluators were asked to take a small piece of injera sample twice and score the taste, and then new samples were provided for mouth feeling, texture, top surface gas holes (size and distribution), color, suppleness, and overall ranking (as an independent score, not an average of the others). A bottle of natural spring water was provided to each evaluator for rinsing his or her mouth after scoring each sample.

Significant differences in sensory injera quality were observed where at least two barley varieties (Haftusene and Himblil, released in 2011) were found to have injera quality equal to tef. Partial least squares regression was used to build models to predict injera sensory quality from pasting properties. These models allowed the separation of Haftusene and Himblil from varieties with lower quality. To investigate if the high injera quality of Himblil was heritable, it was crossed to the intermediate quality Saesa, and 14 F3:7 families were evaluated. The evaluation suggests transgressive segregation for injera sensory quality and flour properties (Abraha and Abay 2017). Some families matched tef in overall quality over four testing environments. The family S × H-T182, derived from the Saesa × Himblil (S × H) cross and officially released in 2012 as a high injera quality variety, is a major achievement for barley breeding in Ethiopia.

Fig. 7.6 Farmers evaluating injera made from different barley varieties and tef at Atsbi Habes, eastern Tigray, in Northern Ethiopia

7.5   Conclusion

Significantly large numbers of rural women-headed households exist in villages of Tigray region. They largely depend upon agriculture that uses little or no purchased inputs in the drought-prone environments. Understandably, farmers prefer to grow their local barley which are well-adapted to their environments than modern and released varieties nationally. As a women plant breeder, I led the team of Mekelle University to assess the needs of these women farmers in barley varieties. I have long-standing research and have led over 15 international projects, and, presently, I am Vice President of the university where I still continue working with women farmers. As a first step, my team supplied farmers seeds of varieties that met their criteria for testing and selection in a participatory varietal selection (PVS) program. Farmers invariably preferred locally developed varieties by farmers, i.e., Himblil. On this, we built a participatory barley breeding (PPB) in a decentralized mode and developed six varieties in partnership with women farmers. These varieties are drought tolerant, early to mature and highly nutritious for food and food products. This has increased productivity of the region from 2 to 5 tons per hectare and has benefitted over 30,000 farmers by increasing their income by about 30–50%. In this chapter, I have discussed the scientific basis of PPB and PVS and their relevance for location-specific breeding of barley in comparison to breeding for wider adaptation in the national program targeted to high-input conditions. I have described several examples of the role of women participation in Tigray in terms of PPB and PVS, seed production and dissemination, increase of biodiversity and differential adoption rates of preferred barley varieties for their uses, and varietal choices for novel food products and finally upgrading women's post-harvest processing and value addition to food chains for evaluating their business prospects.

Being a Tigrain woman myself, I feel satisfied with the success of projects that I have led for the benefit of women farmers of the region. The development and selection of farmer-preferred varieties for various traits was chronologically and scientifi- cally followed from testing, adoption, and seed dissemination to managing post-harvest losses and food systems for business. Our research has shown that national plant breeding policies need to prioritize more efforts on regional plant breeding with emphasis on location-specific variety development, release, seed pro- duction, and food processing. Diversity of food products made by Tigrain women makes them suitable for all ages from infants to older and frail people. This means the government needs to invest more on barley breeding and developing barley grain-based food industry in Tigray region both for local consumption in Ethiopia and for addressing the needs of international markets.

Acknowledgment I would like to acknowledge my Supervisor Prof Asmund Bjørnstad, NMBU, for his selfless dedication to my academic/research development. Dr. Ann Waters Bayer is my men- tor and a friend, from whom I have learned the vital skill of disciplined critical thinking. I am grate- ful for the support of the NUFU (Norwegian seed safety and food science projects), ISSD, SOLIBAM, IDRC, and the collaboration of Kansas State University/USAID-funded Feed the Future Innovation lab for the Reduction of Post-Harvest Loss—Ethiopia.

References

Abay, F., & Bjørnstad, A. (2008). Participatory varietal selection of barley in the highlands of Tigray in Northern Ethiopia. In M. H. Thijssen, Z. Bishaw, A. Beshir, & W. S. de Boef (Eds.), *Farmers, seeds and varieties: Supporting informal seed supply in Ethiopia* (p. 348). Wageningen: Wageningen International.

Abay, F., & Bjørnstad, Å. (2009). Specific adaptation of barley varieties in different locations in Ethiopia. *Euphytica, 167*(2), 181–195. https://doi.org/10.1007/s10681-008-9858-3.

Abay, F., & Halefom, K. (2012). Impact of Participatory Varietal Selection (PVS) on varietal diver- sification and seed dissemination in the tigray region, North Ethiopia: A case of barley. *Journal of the Drylands, 5*(1), 396–401.

Abay F., & Sarah (2009). *Upgrading women's food production value chains for greater price share: Tigray region, Northern Ethiopia- a baseline survey*. IDRC.

Abay, F., Waters-Bayer, A., & Bjørnstad, A. S. (2008, June). *Farmers' seed management and inno- vation in varietal selection: Implications for Barley Breeding in Tigray, Northern Ethiopia* (Vol. 37, No. 4). Ambio: Royal Swedish Academy of Sciences 2008.

Abraha, A., & Abay, F. (2017). Effect of different Cereal Blends on the quality of injera a Staple food in the highlands of Ethiopia. *Momona Ethiopian Journal of Science (MEJS), 9*(2), 231– 241. CCNCS, Mekelle University, ISSN:2220-184X. https://doi.org/10.4314/mejs.v9i2.7.

Abraha, A., Uhlen, A. K., Abay, F., Sahlstrøm, S., & Bjørnstad, Å. (2013). Genetic variation in barley enables a high quality Injera, the Ethiopian staple flat bread, comparable to Tef. *Crop Science, 53*, 1–11. https://doi.org/10.2135/cropsci2012.11.0623.

Annicchiarico, P., Bellah, A., & Chiari, T. (2005). Defining subregions and estimating benefits for a specific adaptation strategy by breeding programs: A case study. *Crop Science, 45*, 1741–1749.

Ceccarelli, S., Grando, S., & Hamblin, J. (1992). Relationships between barley grain yield mea- sured in low and high yielding environments. *Euphytica, 64*, 49–58.

Ceccarelli, S., Grando, S., & Impiglia, A. (1998). Choice of selection strategy in breeding barely for stress environments. *Euphytica, 103*, 307–318.

Ceccarelli, S., Grando, S., Bailey, E., Amri, A., El-Felah, M., Nassif, F., Rezgui, S., & Yahyaoui, A. (2001). Farmer participation in barley breeding in Syria, Morocco and Tunisia. *Euphytica, 122*, 521–536.

Chambers, R. (1997). Learning to learn. In *Whose reality counts? Putting the first last*. London: Intermediate Technology Publications.

Chambers, R., & Ghildyal, B. P. (1985). Agricultural research for resource-poor farmers: The farmer-first-and-last model. *Agricultural Administration and Extension, 20*, 1–30.

Gyawali, S., Joshi, K. D., & Witcombe, J. R. (2002, March 12–15). Participatory plant breeding in rice in low-altitude production systems in Nepal. In J. R. Witcombe L. B. Parr & G. R. Atlin (Eds.), *Breeding rainfed rice for Drought-prone environments: Integrating conventional and participatory plant breeding in South and Southeast Asia*. Proceedings of a DFID Plant Sciences Research Programme/IRRI Conference (pp. 79–89). Bangor/Manila: Centre for Arid Zone Studies (CAZS)/International Rice Research Institute (IRRI)

Joshi, A., & Witcombe, J. R. (1996). Farmer participatory crop improvement. II. Participatory varietal selection, a case study in India. *Experimental Agriculture. 32*: 461–477.

Joshi, K. D., Sthapit, B. R., Subedi, M., & Witcombe, J. R. (2002). Participatory plant breeding in rice. In D. A. Cleveland & D. Soleri (Eds.), *Farmers, scientists and plant breeding: Integrating knowledge and practice* (pp. 239–267). Wallingford: CABI.

Meehan, F. (2004, December). *Female headed household in Tigray, Ethiopia: A study review* (Dryland Coordination Group (DCG) Report No. 35), pp. 1–35.

Morris, M. L., & Bellon, M. R. (2004). Participatory plant breeding research: Opportunities and challenges for the international crop improvement system. *Euphytica, 136*, 21–35.

Okali, C., Sumberg, J., & Farrington, J. (1994). *Farmer participatory research. Rhetoric and reality*. London: Intermediate Technology Publications.

Paris, T. R., Singh, A., Cueno, A. D. & Singh, V. N. (2008). Assessing the impact of participatory research in rice breeding on women farmers: a case study in eastern Uttar Pradesh, India. *Experimental Agriculture, 44*, 97–112.

Petros, S., Abay, F., Desta, G., & O'Brien, C. (2018). Women farmers' (dis)empowerment compared to men farmers in Ethiopia. *World Medical & Health Policy, 10*(3), 220–245.

Rhoades, R. E., & Booth, R. H. (1982). Farmer-back-farmer: A model for generating acceptable agricultural technology. *Agricultural Administration, 11*, 127–137.

Snapp, S. (1999). Mother and baby trials: A novel trial design being tried out in Malawi. *Target Newsletter of the Southern Africa Soil Fertility Network, 17*, 8–10.

Sperling, L., & Berkowitz, P. (1994). *Partners in selection: Bean breeders and women bean experts in Rwanda*. Washington, DC: CGIAR Gender Program/Consultative Group on international Agricultura! Research.

Sperling, L., Loevinsohn, M. E., & Ntabomvura, B. (1993). Rethinking the farmers' role in plant breeding: Local bean experts and on-station selection in Rwanda. *Experimental Agriculture, 29*, 509–519.

Sperling, L., Ashby, J. A., Smith, M. E., Weltzien, E., & McGuire, S. (2001). A framework for analyzing participatory plant breeding approaches and results. *Euphytica, 122*, 439–450.

Sthapit, B. R., Joshi, K. D., & Witcombe, J. R. (1996). Farmer participatory crop improvement. III. Participatory plant breeding. A case study for rice in Nepal. *Experimental Agriculture, 32*, 479–496.

Sulaiman, R. V., & Hall, A. (2002). Beyond technology dissemination: Reinventing agricultural extension. *Outlook on Agriculture, 31*, 225–233.

Virk, D. S., Singh, D. N., Kumar, R., Prasad, S. C., Gangwar, J. S., & Witcombe, J. R. (2003). Collaborative and consultative participatory plant breeding or rice for the rainfed uplands of eastern India. *Euphytica, 132*, 95–108.

Virk, D. S., Chakraborty, M., Ghosh, J., Prasad, S. C., & Witcombe, J. R. (2005). Increasing the client-orientation of maize breeding using farmer participation in eastern India. *Experimental Agriculture*. in press.

Weltzein, E., Smith, M., Meitzner, L. S., & Sperling, L. (2003). Technical and institutional issues in participatory plant breeding from the perspective of formal plant breeding. A global analysis of issues, results and current experiences. In: PPB Monograph No. 1, 1–208. CGIAR Systemwide Program on Participatory Research and Gender Analysis for Technology Development and Institutional Innovation: Centro International de Agricultural Tropical (CIAT).

Witcombe, J. R., & Virk, D. S. (2001). Number of crosses and population size for participatory and classical plant breeding. *Euphytica, 122*, 451–462.

Witcombe, J. R., Joshi, A., Joshi, K. D., & Sthapit, B. R. (1996). Farmer participatory crop improvement. I. Varietal selection and breeding methods and their impact on biodiversity. *Experimental Agriculture, 32*, 445–460.

Witcombe, J. R., Joshi, A., & Goyal, S. N. (2003). Participatory plant breeding in maize: A case study from Gujarat, India. *Euphytica, 130*, 413–422.

Witcombe, J. R., Joshi, K. D., Gyawali, S., Musa, A. M., Johansen, C., Virk, D. S., & Sthapit, B. R. (2005). Participatory plant breeding is better described as highly client-oriented plant breeding. I. Four indicators of client-orientation in plant breeding. *Experimental Agriculture, 41*, 299–320.

Witcombe, J. R., Gyawali, S., Subedi, M., Virk, D. S., & Joshi, K. D. (2013). Plant breeding can be made more efficient by having fewer, better crosses. *BMC Plant Biology, 13*, 22. https://doi.org/10.1186/1471-2229-13-22.

Yan, W., Kang, M. S., Ma, B., Woods, S., & Cornelius, P. L. (2007). GGE biplot vs. AMMI analysis of genotype by-environment data. *Crop Science, 47*, 641–653.

Dr. Fetien Abay Abera (PhD) is currently the vice president for research and community services, Mekelle University, Ethiopia. Dr. Fetien has a diploma in plant science from Hawassa College of Agriculture (1985) and a bachelor of science (1991) from Haramaya University of Agriculture, Ethiopia. She received her master's in Rural Resource Management (RRM) from University of Wales, Bangor, UK in 1996, and a PhD in plant breeding and seed science from the Norwegian University of Life Sciences in 2007.

Dr. Fetien is the first female professor in plant breeding and seed in Ethiopia. She has done an outstanding research as evidenced by the release of six climate resilient, high yielding and quality food and malt barley varieties. As the Integrated Seed Sector Development, Tigray region coordinator since 2009, she organised 50 seed producer groups who are currently supplying the major seed requirements for the Tigray region. She is also known for her ability to win large projects and creating synergy among them which has resulted in high impact of projects on community development. For example, the Norwegian Supported Seed Safety through diversity, Women and Food Science: together towards national visibility and IDRC supported project on"Upgrading Women's Food Product Value Chain" were examples of those many projects she led. Her involvement as the national coordinator on a five year project (2014–2018) on USAID's Feed the Future Innovation Lab on Postharvest Loss in collaboration with Kansas State University (USA) has benefitted farmers and five PhD students at both Mekelle and Bahir Dar Universities. She has released 6 climate resilient, high yielding, high quality barley varieties. She has published 85 publications (35 Journal and 50 popular journals, book chapters, peer reviewed proceedings) in her career. Award & Recognitions Professor Fetien has received numerous awards and recognitions, but her greatest asset is the support she has provided to women in agriculture to bring awareness about gender equality, an often ignored aspect of the role of women in agriculture. She is a dedicated university professor who has supervised over 60 MSc/PhD graduate students. In addition, she is actively involved in upgrading women Diploma students to BSc level (as she herself did many years ago). In 2009, at the African wide Women Professionals in Science Competition, she was recognized among the top five African women scientists for her work in plant breeding, and the Ethiopian Government honored her outstanding contributions to linking science and farmers and to furthering women's development with the President's Award in 2010. Professor Fetien also received the "East African Laureates" prestigious prize of the African Union Kwame Nkrumah Prestigious Scientific achievement Award, in 2014. Her selection for this prestigious award was based on her academic and scientific publications (number of peer reviewed journals, book chapters), number of inventions and/or technologies registered, research impact for Africa's socio-economic development, supervision of African students, and contribution to African Libraries. In 2017, Fetien emerged as the second runner-up at the prestigious Impact Research and Science in Africa (IMPRESSA) Awards. The lives and careers of women like Professor Abera speak to the wealth of under-recognized female talent in Africa's agricultural research sector.

Chapter 8
Synthesis of a Research Program in Precision Poultry Environmental Control Using Biotelemetry

Takoi Khemais Hamrita, Taylor Ogle, and Amanda Yi

Contents

To me, precision agriculture is about listening—listening to the needs of the land, the animals, and the environment. It's about being holistic and tuning in to all stakeholders and parts of a system to make informed optimal decisions. It is no surprise that when I joined the department of biological and agricultural engineering at the University of Georgia (UGA), and I determined that I was going to apply my electrical engineering and systems background to developing a research program that revolved around environmental control for poultry production, the first thing I thought about was, what do the birds tell us?

T. K. Hamrita (✉) · T. Ogle · A. Yi
School of Electrical and Computer Engineering, College of Engineering,
University of Georgia, Athens, GA, USA
e-mail: thamrita@uga.edu

© Springer Nature Switzerland AG 2021
T. K. Hamrita (ed.), *Women in Precision Agriculture*, Women in Engineering
and Science, https://doi.org/10.1007/978-3-030-49244-1_8

In 1995, I took an assistant professor position at UGA in the department of biological and agricultural engineering. This was a bold move considering that (1) I had no background in agriculture, (2) my background was almost all theoretical, and (3) the position was very applied and revolved around applying my controls and systems background to an area of agriculture. Being in Georgia, I quickly learned that poultry production was very important to the state and chose to focus my research program in this area.

When I started researching poultry production, I quickly learned that:

1. "Modern" birds, much larger in size than their ancestors (growth rate of about 350% compared to that of the 1957 chicken due to improved genetics and nutrition, (Havenstein et al. 1991)), grown in confined facilities, aren't as hardy as birds grown years ago (Czarick and Lacy 1994). Today's birds seem to have a lesser ability to cope with less than optimal environmental conditions (Hamrita and Mitchell 1999).

2. Environmental stressors cause substantial losses to poultry growers and the poultry industry.

3. Despite availability of sophisticated control technology, poultry environmental control was very basic and often aimed at maintaining temperature in the environment within a set point by controlling ventilation and heating rates (Lamade 1984; Mitchell 1984; Worley and Allison 1984). The control actions were often based on feedback measurements of ambient temperature collected from a single location in the building (Aerts et al. 1996).

4. New US regulations on reduction of disease pathogens in consumer products (Moriarty1993) required more effort to be put into management, biosecurity, and control of the environment, all of which would necessitate the adoption of more advanced control methodologies (Hamrita and Mitchell 1999).

5. My systems thinking dictated that I look at the poultry-growing facility as a system, the most important part of which is the birds. Current environmental control practices failed to factor in the bidirectional interaction of the poultry with the environment, therefore missing out on feedback from the most important part of that system, the birds.

6. Unlike mechanical or electrical systems where it's commonplace to sense important variables, model the behavior of the system, and design dynamic controllers that met specific objectives, poultry environmental control presented some unique challenges.

   • Sensing important physiological and behavioral feedback from the birds without interfering with their normal behavior was a challenge. Biotelemetric sensors for doing so were very uncommon and pricey, and studies for their use were rare.
   • Birds are complex living beings, and models for how they respond to environmental stimuli were not available. This is due to the difficulty of obtaining high-quality data and the lack of fundamental research to evaluate the impact of environmental factors on birds' physiology.

- Precedence for a control system that adjusted environmental conditions in real time based on physiological needs of the birds did not exist.

The above factors presented intellectually stimulating research prospects for me: (1) I get to work with living beings, (2) I get to work with state-of-the-art biotelemetry sensors, and (3) I get to work with hardware, design experiments, collect data, develop models, and apply my theoretical knowledge to a real-life situation. Most of all, the idea of a chicken "talking" to a controller to tell it how to adjust environmental conditions to keep the chicken comfortable was very stimulating to me and presented the opportunity for me to be a pioneer in developing new understanding that could lead to more optimal environmental controllers for poultry production, therefore alleviating stress and improving the economic bottom line for growers.

This chapter highlights the significant milestones achieved by my research program towards using biotelemetry to build precision poultry environmental controllers that respond directly and in real time to the needs of the birds. The discussion is presented in three sections: The first section focuses on biotelemetry and its use to monitor poultry deep body temperature (DBT) responses to various environmental conditions, the second section deals with DBT modeling efforts to date, and the third section presents results of the first poultry environmental controller prototype which responds to poultry DBT responses in real time.

8.1 Biotelemetry and Monitoring Poultry DBT Responses to Environmental Stressors

Environmental stresses cause significant economic losses due to increased mortality, downgrading, and condemnations of carcasses and associated problems of environmental pollution, reduced production, reduced feed intake and body weight gain, and impaired immune function (Payne 1966, as cited in Green and Xin 2009; Mader et al. 2002, Brown-Brandl et al. 2003, and Hahn 1997 all as cited in Silva et al. 2005, Hamrita and Paulishen 2011). Heat stress develops when birds' ability to thermoregulate and maintain homeostasis under stressful ambient conditions fails. In years past, farmers have relied heavily on experience and observations in order to be able to tell when environmental stressors start to negatively impact the animals. By the time the farmers could notice any changes in the animals, it was too late to do anything about it (Hamrita et al. 1998). More reliable methods for monitoring animal responses to environmental stressors have been addressed in the literature, including monitoring of feed consumption using weighing feeders (May and Lott 1992; Korthals et al. 1992), monitoring of behavioral responses using video cameras (Korthals et al. 1992), monitoring of generated heat using calorimetric methods (Korthals et al. 1992; Schurmas et al. 1996), and monitoring of animal weight (DeShazer and Randall 1988). These methods, although more reliable than farmer observations, still have the drawback of not detecting stress at its earliest stages (Hamrita et al. 1998). Noticeable responses to environmental stressors are often preceded by internal physiological responses which are usually the first stress indicators, such as a change in deep body temperature and/or heart rate (Hamrita et al. 1998; Mitchell 1981). If measured properly, these responses are the ultimate indicators of stress and allow detection of stress at much earlier stages (Hamrita et al. 1998). Monitoring and understanding animal physiological responses to environmental stressors is a necessity if we are to design precision management systems that respond to the needs of the animals. A less stressful environment would ultimately lead to the production of healthier animals (Hamrita et al. 1998).

Our early research efforts have focused on (1) testing and validating new commercially available telemetry systems and measurement techniques to demonstrate their effectiveness for accurate continuous monitoring of poultry DBT, (2) monitoring and evaluating DBT responses of poultry under various stressful environmental stimuli, (3) gaining a better understanding of poultry thermoregulatory responses, and (4) evaluating and establishing DBT as a viable indicator of heat stress.

8.1.1 Biotelemetry (Based on Hamrita and Paulishen 2011)

Biotelemetry is the remote detection and measurement of physiological, bioelectrical, and behavioral variables to monitor function, activity, or condition of conscious unrestrained humans or animals. This reduces stress and physiological disturbance by removing the influence of the measurement procedure and thereby improving the quality of data. Biotelemetry encompasses a broad range of techniques of varying invasiveness including radio tracking and the use of internally or externally mounted remote sensing systems (Morton et al. 2003). The output of the sensor is transmitted wirelessly from the animal to a receiver. A data acquisition system extracts the measured variable through proper signal conditioning and calibration.

Biotelemetry provides the opportunity to increase the frequency of observation or continuously monitor several variables over extended periods of time, therefore providing increased access to larger amounts of physiological data (Baras and Lagardère 1995). Additionally, biotelemetry makes possible real-time processing of data and the ability to act on it. Knowing how key variables are changing in real time in animals allows, for instance, faster adjustment of feeding times to activity rhythms, more objective identification of the preference/tolerance margins toward environmental variables, and precise assessment of the impact of environmental or operational changes (Baras and Lagardère 1995).

The use of biotelemetry by biologists and veterinarians for physiological and biomedical research is quite common; however, the use of this technology in farming practices is still a novel undertaking. Biotelemetric measurement techniques which make it possible to easily and economically monitor physiological responses in production animals without constraining the animal and attaching it to several wires and probes and without interfering with the animals' normal daily routines must be determined and validated (Hamrita et al. 1998).

8.1.2 DBT Responses to Stressful Ambient Temperature and Relative Humidity Conditions (Based on Hamrita et al. 1997 and Lacey et al. 2000a)

Poultry are homeothermic animals. So long as environmental conditions are within their thermoneutral zone, they are capable of maintaining a core DBT between 41.2 °C and 42.2 °C. Under stressful heat conditions, biophysical defense mechanisms break down leading to elevated DBT responses, which could potentially result in death, causing economic loss (Tau and Xin 2003, as cited in Hamrita and Conway 2017a, b).

In Hamrita et al. (1997), our first study, we evaluated the use of a biotelemetry system (Minimitter, Bend, Oregon; Telonics, Inc., Mesa, Arizona) to monitor DBT of poultry, using implanted sensors, under various ambient temperature (AT) conditions. Figure 8.1 shows an example of DBT response of a bird to an 11 degree step change in ambient temperature from 21 °C to 32 °C. For the first time, we were able to witness the bird's thermoregulatory system in action as it tried to cope with the temperature increase and as it worked to lower the bird's DBT.

In Lacey et al. (2000a), we assembled a more advanced telemetry system using smaller implantable sensors by AVM (AVM Instrument Company, LTD., Livermore, CA), a multichannel programmable receiver by Telonics (Telonics Inc., Mesa, AZ), and a data acquisition system by Minimitter (Minimitter, Bend, Oregon). The goal

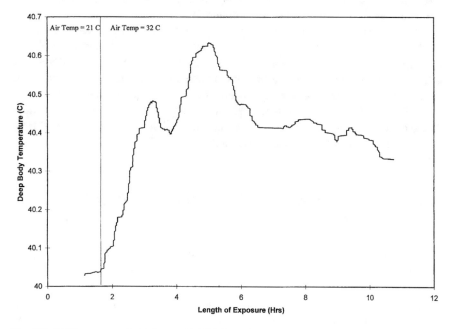

Fig. 8.1 DBT response of a bird to an 11 °C AT step increase (Hamrita et al. 1998)

was to determine the effects of stressful ambient temperature and relative humidity conditions on poultry DBT.

As cited in Hamrita and Paulishen (2011), a high level of relative humidity (RH) is commonly known to be an exacerbating factor in poultry heat stress problems (Brown-Brandl et al. 1997; Lacey et al. 2000a). However, as indicated in Shlomo et al. (1995) and Lacey et al. (2000a), its exact effects have not been "clearly eluci-dated." Hence, more research efforts are necessary to better understand the com-bined effects of AT and RH on poultry and to incorporate this knowledge in optimizing poultry housing management and control. Information on the interactive effects of AT and RH on poultry subjected to heat stress is meager (Yanagi et al. 2002). Humidity can aggravate the adverse effects of high temperature (Steinbach 1971, as cited in Tao and Xin 2003a) because animals increasingly rely on latent heat loss with rising temperature (Tao and Xin 2003a).

In our study, three levels of AT (31 °C, 34 °C, and 37 °C) and two levels of RH (50% and 80%) were considered. Figure 8.2 shows an example of mean DBT responses for 12 birds to treatments with increasing AT at both RH levels. This fig-ure indicates that the effects of AT and RH on mean deep body temperature of broil-ers are cumulative. Higher RH increases the effective AT experienced by the bird and results in higher DBT (Lacey et al. 2000a). Notice, for instance, the similarity in responses to 37 °C AT at 50% RH in Fig. 8.2a and 31 °C AT at 80% RH in Fig. 8.2c. A 6 °C increase at 50% RH in Fig. 8.2a had the same effect as a 30% in RH at 31 °C in Fig. 8.2c. This is an important result that allows for better under-standing of how environmental conditions have compounded effects on the physiol-ogy of the birds. Environmental controllers that only take into consideration ambient temperatures are ignoring half of the problem.

Overall, results in both of the above studies confirmed that (1) measurements of DBT responses to ambient temperature and relative humidity stressors were consis-tent among all birds, (2) responses were noticeably and significantly different for different environmental conditions, and (3) changes in responses from one set of conditions to the other was clearly attributed to the change in ambient conditions and not to fluctuations in the measurement system or in between bird variation. These results are very significant as they make it possible to use DBT as a real-time indicator of poultry heat stress (Hamrita and Paulishen 2011).

8.1.3 DBT Responses to Air Velocity (Based on Hamrita and Conway 2017b)

Tunnel ventilation is the method of choice within the poultry industry to alleviate heat stress in poultry. Setting air velocity within poultry chambers to levels that ensure bird comfort while optimizing performance is an important goal. Historically, simple thermostats measuring ambient temperature have been used to indicate when to turn on the fans, for how long, and at what speeds (Simmons et al.1993). In this

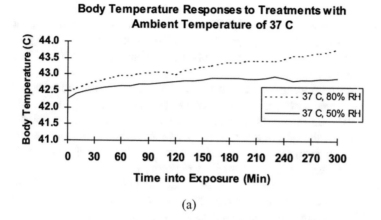

(a)

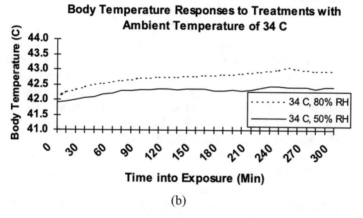

(b)

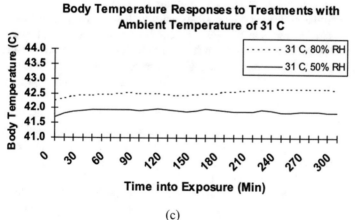

(c)

Fig. 8.2 Mean DBT responses of 12 birds to different RH at three different ATs: (**a**) 37 °C at 80% RH and 50% RH, (**b**) 34 °C at 80% RH and 50% RH, and (**c**) 31 °C at 80% RH and 50% RH (Lacey et al. 2000a)

study, we presented an approach for evaluating the effects of tunnel ventilation on poultry deep body temperature (DBT) using biotelemetry. In particular, the study aimed to answer the following questions: (1) Does DBT rise quickly, significantly, and consistently as a result of increased air temperature? (2) Does tunnel ventilation alleviate heat stress and lower DBT quickly and consistently? The study also reported on birds' weight gain with and without tunnel ventilation. Heat-stressed birds were subjected to high-velocity tunnel ventilation to evaluate the impact, if any, on birds' DBT responses. Three consecutive experiments were conducted using six birds each at the ages of 8.6, 9.0, and 9.4 weeks. Birds were subjected to stressful step changes in ambient temperature, provoking drastic DBT increases, in order to evaluate the effectiveness of tunnel ventilation in lowering the DBT response. Experiments spanned approximately 12 h each and led to 18 data sets. DBTs of the birds began to rise almost immediately after application of the step temperature and continued to rise until ambient temperature was brought down.

Figure 8.3 shows an example of average group DBT responses of heat-stressed birds of different ages under tunnel ventilation (treatment birds) compared with average responses of control birds subjected to no ventilation. The figures indicate that the impact of air velocity is almost immediate. Under tunnel ventilation, DBTs rose as a result of the step increase in ambient temperature; however, compared to control birds, they did at a slower rate and achieved lower maxima during the same period. The minimum DBT increase for treatment birds in all three experiments was 0.7 °C, compared to 1.5 °C for control birds; the maximum DBT increase was 1.9 °C compared to 3.0 °C for control birds. Group average DBT increases for each experiment ranged between 0.5 °C and 1 °C compared to 1–1.9 °C for control birds. DBT responses of treatment birds were consistently lower than those of the control group.

During experiment days, birds exposed to tunnel ventilation consistently gained weight with a percentage weight gain ranging from 1% to 11%. Birds not exposed to tunnel ventilation behaved less consistently with some gaining as much as 14% while others lost as much as 9%. Pearson correlation and scatter plots were used to determine that DBT responses of birds to step increases in air velocity were for the most part consistent with one another. One important application of this latter result is that a small group of birds' DBTs could be used to estimate the response of the flock.

Although further studies are required to derive more comprehensive and more statistically significant results, this study provided preliminary data that is needed to warrant such studies and a stepping-stone for making optimal management and risk assessment decisions that are based on physiological needs of the birds. Overall, the study strengthened the notion that DBT rises quickly, significantly, and consistently as a result of heat stress and that increased airflow mitigates the problem quickly and consistently. These results lay a stronger foundation for designing tunnel ventilation practices that optimize the well-being of the birds.

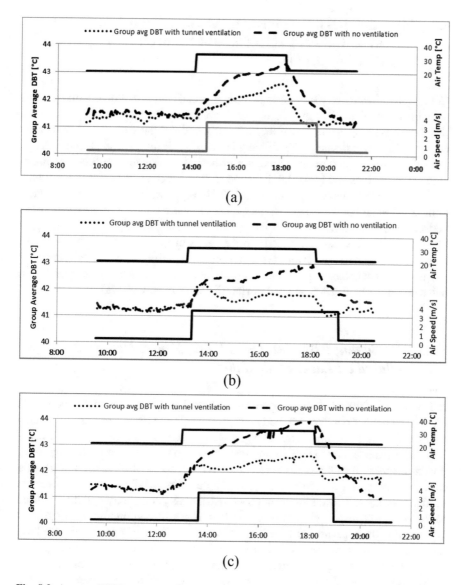

Fig. 8.3 Average DBT responses of control and treatment birds to tunnel ventilation: (**a**) birds 8.6 weeks old, (**b**) birds 9.0 weeks old, and (**c**) birds 9.4 weeks old (Hamrita and Conway 2017b)

8.2   Modeling DBT Responses to Ambient Temperature (Based on Lacey et al. 2000c and Hamrita and Conway 2017a)

Continuous biotelemetry monitoring of poultry makes it possible to gain dynamic responses that relate birds' physiology with environmental factors. Deep body temperature (DBT) has been shown through our work and the work of others (Lacey et al. 2000a, b; Aengwanich 2008; Yang et al. 2007; Hamrita et al. 1998) to be an immediate, strong, and reliable indicator of heat stress. Using real-time records of ambient conditions and birds' DBT responses to these conditions makes it possible to develop models that can predict birds' DBT responses to future ambient conditions. Studies and models that help us gain better understanding of the relationship between DBT and environmental factors are necessary to help us make management decisions that respond directly to the needs of the birds. For instance, these models form the basis of various types of dynamic controllers, where the controller has to predict responses to future control actions to optimize current and future behavior (Aerts et al. 2003 as cited in Hamrita and Conway 2017a).

8.2.1   Predicting DBT Responses Using Neural Networks (Based on Lacey et al. 2000c)

In Lacey et al. (2000c), we developed and evaluated artificial neural network models (ANNs) to predict poultry deep body temperature (DBT) responses to stressful step changes in ambient temperature. The goal was to examine the generalization capability of the models by (1) predicting responses of birds not used in training to AT conditions used in training, (2) predicting DBT responses of birds used in training to AT conditions not used in training, and (3) predicting DBT responses of birds not used in training to AT conditions not used in training. A telemetry system was used to measure DBT responses of birds under various heat stress conditions. Three birds were used in the experiment under five different heat stress schedules. The resulting data was used to train and test various neural network architectures. Through trial and error, a recurrent Elman-Jordan network was determined to be the most suitable architecture. Figure 8.4 shows an example of the predicted versus measured DBT responses together with the corresponding AT schedule. Predictions were for an ambient temperature schedule not used to train the network. The mean absolute error was 0.081 °C and R^2 was 0.94. The agreement between the predicted and measured DBT is evident throughout the duration of the experiment. Note that not only did the model predict an increase in DBT with AT increase, but it also predicted a decrease in DBT with AT decrease.

In general, the model's performance was (1) reasonably well when predicting responses of a different bird to AT schedules used in training, (2) well when predicting responses of a bird used in training to new AT schedules, and (3) not as well

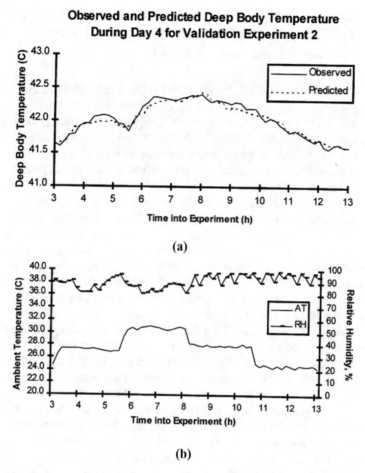

Fig. 8.4 An example of predicted versus measured DBT responses (**a**) to stressful step changes in ambient temperature (**b**) together with the corresponding AT schedule. Predictions were for an ambient temperature schedule not used to train the network. The mean absolute error was 0.081 °C and R^2 was 0.94 (Lacey et al. 2000c)

when predicting DBT responses of a bird not used in training to AT conditions not used in training. This is not surprising considering the small data set used in training. Using larger data sets with more birds and more AT schedules would likely lead to improved DBT predictions. Results of this study indicated that using neural networks is a promising approach for predicting DBT responses to stressful AT conditions. Future research efforts should focus on incorporating larger data sets from a bigger sample of birds with a more extensive and comprehensive set of environmental conditions.

8.2.2   First-Order Dynamics: A Good Approximation of Poultry DBT Responses to Ambient Temperature (Based on Hamrita and Conway 2017a)

In Hamrita and Conway (2017a, b), we developed an empirical approach for identifying the order of dynamics that govern the relationship between DBT and ambient temperature. Forty-six DBT/AT data sets involving three birds and DBT responses to 23 upward AT steps, and 23 downward AT steps were obtained using a biotelemetry system. For the sake of generating the model, ambient steps were chosen high enough to push the chickens out of their homeothermal zone and to produce a very noticeable measurable response.

DBT rose almost immediately as the upward step was applied and started decreasing almost just as quickly as the downward step was applied. One thing that was strongly evident from the plots is that regardless of environmental conditions, the DBT response trends among the chickens were consistent. This is in agreement with results in several previous studies, which have arrived at similar conclusions (Hamrita and Hoffacker 2008; Lacey et al. 2000a; Hamrita et al. 1998). Examination of DBT responses under the AT test conditions confirmed that DBT responses of each heat-stressed bird varied dynamically, predictably, and measurably to AT pulses. Additionally, responses of the three birds were consistent with one another (0.88 average Pearson correlation coefficient, $p < 0.0005$). Scatter plots were developed to study the relationship between birds' responses to the same environmental conditions, and the results suggested a fairly linear relationship.

The data indicated that DBT responses to AT followed a first-order behavior in most cases with an average time constant of 1.6 h, and the curve fitting method was used to validate this observation. The time constant, initial temperature, and final temperature constants in the assumed first-order model were determined to best fit the experimental data. There was a 0.88 average correlation between DBT model and measured data ($p < 0.0005$). Figure 8.5 shows an example average DBT response for one of the experiments overlaid with the theoretical response using the first-order exponential model. The average response was calculated by averaging the responses of the three birds for that experiment.

Although chickens are complex living beings with complex cooling systems, Fig. 8.5 and the correlation analysis seem to indicate that a first-order model is a decent approximation. Notice that the model led to a closer approximation when the steps were going upward. Measured downward DBT responses were, almost always, faster to go down than the model response. This result implies that the dynamic DBT response to step changes in AT is asymmetrical and is faster when AT (therefore DBT) decreases, and it is slower when AT (therefore DBT) increases. This asymmetry was observed in a study by Aerts et al. (2003) in which the authors modeled heat production responses to step changes in ambient temperature using a first-order model wherein responses to downward steps (increased heat production) were faster than those to upward steps (decreased heat production).

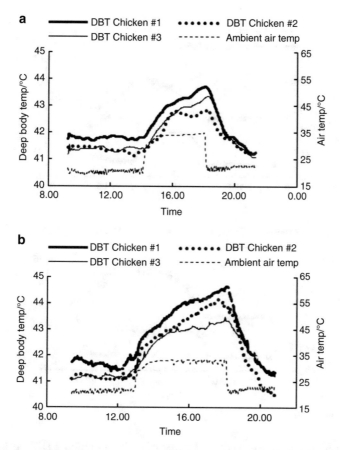

Fig. 8.5 Example average DBT response overlaid with the theoretical response using the first-order exponential model. (**a**) Experiment #1, DBTs of chickens 8.6 weeks old. (**b**) Experiment #3, DBTs of chickens 9.4 weeks old. (**c**) Experiment #1, 8.6 weeks old group average DBT versus model. (**d**) Experiment #3, 9.4 weeks old group average DBT versus model (Hamrita and Conway 2017a)

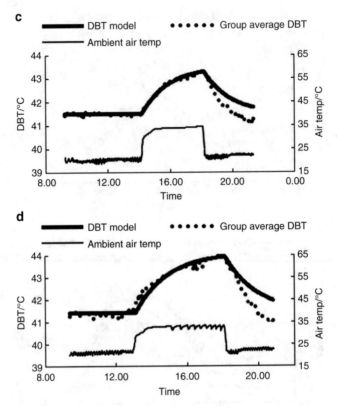

Fig. 8.5 (continued)

To our knowledge, this study is the first of its kind to model DBT as a function of AT and to suggest that a first-order dynamic model represents a good approximation of poultry dynamic DBT responses to large step changes in AT. Although further studies are needed to more fully derive the model, this study provided a stepping-stone toward gaining a better understanding of the relationship between DBT and AT, therefore taking us one step closer toward making optimal management that are based on physiological needs of the birds.

8.3 Closed Loop Control of Poultry DBT Using Variable Air Velocity (Based on Hamrita and Hoffacker 2008)

Environmental control is essential in the alleviation of heat stress in poultry environments. The most common and basic form of control in studies reported in the literature aims to maintain temperature around a set point by controlling tunnel ventilation and heating rates. In most cases, the controller is based on feedback measurements of ambient temperature collected from a single location in the building using a thermistor or a thermocouple (Aerts et al. 1996; Hamrita and Mitchell 1999). Although temperature remains the most widely studied variable, some advanced studies aimed to control other environmental variables such as humidity, static pressure, and ventilation rates (Timmons et al. 1995; Mitchell 1986, 1993; Allison et al. 1991; Timmons and Gates 1987; Flood et al. 1991; Zhang 1993; Geers et al. 1984; Berckmans and Goedseels 1986; Hamrita and Mitchell 1999). Our research (Hamrita and Mitchell 1999; Hamrita et al. 1997; Hamrita et al. 1998; Lacey et al. 2000a, b, c; Hamrita and Hoffacker 2008) and that of a few other investigators (Aerts et al. 1996; Goedseels et al. 1992; Barnett and Hemsworth 1990) suggest that environmental controllers can be improved by gaining insight into physiological responses of the birds to environmental stressors. To our knowledge, our research program is the first of its kind to develop an environmental controller prototype that responds in real time to birds' DBT responses. Prior studies that investigated the relationship between deep body temperature (DBT) and environmental variables (Hamrita et al. 1997, 1998; Lacey et al. 2000a, b, c; Hamrita and Hoffacker 2008) laid the foundation for such a controller. In these studies, we established that DBT is a significant, measurable, effective, and predictable indicator of heat stress in poultry (Hamrita and Paulishen 2011).

Using an experimental tunnel ventilation enclosure placed inside an environmentally controlled chamber, implanted radiotelemetry sensors, and a programmable logic controller, a proportional integral-type feedback controller was designed to maintain poultry DBT, under stressful ambient temperature conditions, below a given set point by controlling air velocity rates (Hamrita and Paulishen 2011). Figure 8.6 shows the experimental setup for this study (Hamrita and Hoffacker

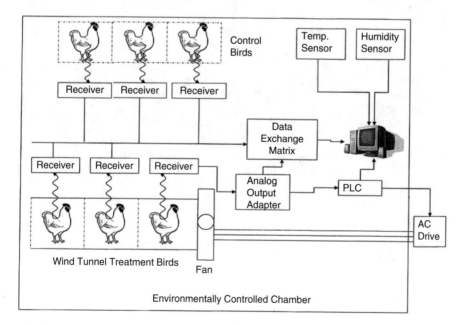

Fig. 8.6 Tunnel ventilation DBT controller experimental setup (Hamrita and Hoffacker 2008)

2008). Three test birds were placed inside the tunnel, and three control birds were placed outside the tunnel. Figure 8.7 shows an example of DBT responses of the three birds inside the tunnel under stressful heat conditions. The top curve shows the variable air velocity produced by the controller to maintain DBT of the birds under a set point. The bottom curves show the birds' DBT responses and how, for the most part, the controller kept these responses below the set point. Figure 8.8 shows the mean DBT responses of the three birds in the tunnel compared to the mean DBT responses of the control birds, for the same experiment as in Fig. 8.7. Although more work is necessary to develop a more stable controller, it is evident that while control birds' DBTs rose significantly as a result of increase in ambient temperature, those of the birds inside the tunnel were kept around the desired set point. Moreover, DBT responses to tunnel ventilation were almost immediate. Throughout the study, the controller maintained the DBT response of the feedback bird below the set point 70%–85% of the time. This study is very encouraging in that it suggested a number of important preliminary results: (1) Air velocity has a measurable, dynamic, and almost immediate impact on DBT of birds under heat stress, (2) DBT of heat-stressed birds can be maintained below a set point by varying air velocity using feedback control, and (3) using DBT as a feedback variable to control air velocity within poultry housing is a promising approach.

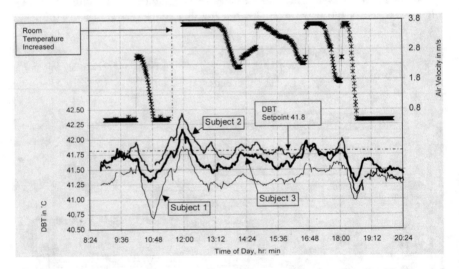

Fig. 8.7 Treatment group DBTs and air velocity generated by the controller (Hamrita and Hoffacker 2008)

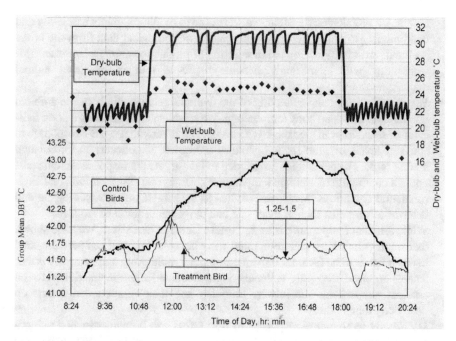

Fig. 8.8 Mean DBT responses for treatment and control birds with and without tunnel ventilation (Hamrita and Hoffacker 2008)

8.4 Conclusion

As the poultry industry strives to produce healthier birds and increase profit margins, they have to face the increasing challenge of providing the birds with environments that are optimal throughout the entire duration of their growth. One way that this challenge could be met is by tapping into birds' physiological responses to environmental stressors to guide environmental management and control decisions. Environmental stressors are unavoidable in poultry production, and these stressors impact the growth and health of the chicken. Most current solutions to controlling environmental stressors rely solely on environmental factors and ignore the physiological needs and responses of the animals (Hamrita and Conway 2017a, b). Methods for monitoring, quantifying, modeling, and predicting animal physiological responses to environment conditions are still at the experimental stage. Methods that use this information to better control the environment are still rare. Our research program in precision control of poultry environments using physiological feedback from the birds has pioneered a number of important results and laid the foundation for a more optimal controller that responds directly and in real time to the dynamic needs of the birds. In particular, we have established the following stepping stones:

- Biotelemetry is a viable tool not only for studying the effects of environmental conditions on chickens but also as a real-time monitoring tool. Using a fully automated biotelemetry system, we were able to monitor animals wirelessly and continuously to obtain accurate high-quality physiological data from the birds. As the interest in physiological monitoring of production animals increases, this technology will increasingly become a viable tool in precision animal management.

- We gained a better understanding of the dynamic relationship between ambient temperature and deep body temperature. We established the latter as a good indicator of heat stress in chickens. We have shown that dynamic DBT responses to stressful levels of ambient temperature are quick, significant, and consistent among birds. Further studies are needed to investigate the responses of chickens under a continuum of ambient temperature changes that closely mimic real environmental conditions. Moreover, studies to investigate other physiological and behavioral variables such as heart rate and movement should be explored.
- While ambient temperature is an important factor in heat stress, we have shown that relative humidity can compound the effects of ambient temperature, and hence it needs to be factored into environmental monitoring and control. Further studies are needed to fully understand and characterize the interaction between ambient temperature and relative humidity and their compounded effects on deep body temperature.
- Deep body temperature responses to heat stress can be approximated using a first-order dynamic model. Model parameters may vary nonlinearly with age and acclamation, and further studies are necessary to better understand and model this complex relationship. Real-time data analysis techniques would have to be investigated to identify and account for data trends that would suggest time-varying, higher-order, or nonlinear dynamics.
- Neural networks present a promising technique to model and predict birds' DBT responses to stressful changes in ambient temperature. These models have the potential of capturing nonlinear dynamics.
- Tunnel ventilation is an important factor in alleviating heat stress in poultry. Our work has shown that tunnel ventilation has immediate cooling effects on poultry DBT responses. Additionally, our prototype controller has shown that air velocity can be controlled dynamically to keep deep body temperature within a desired range.

It is important to note that experiments conducted for our studies involved a small number of birds in a small experimental environmentally controlled chamber. In order to achieve results of higher significance, further experiments would need to be conducted in a real poultry housing environment involving a larger number of birds and more realistic environmental conditions. Also, as we develop more sophisticated environmental controllers for poultry housing, we have to conduct studies that demonstrate to the producers the short-term and long-term economic benefits of these controllers. Producers are generally reluctant to adopt new technologies unless there's demonstrated economic benefit. Without this type of connection to economic benefits, our sophisticated control schemes will not make it to the field.

References

Aengwanich, W. (2008). Effects of high environmental temperature on the body temperature of Thai indigenous, Thai indigenous crossbred and broiler chickens. *Asian Journal of Poultry Science*, 48–52.

Aerts, J. M., Berckmans, D., & Schurmans, B. (1996). On-line measurement of bioresponses for model-based climate control in animal production units. In *Sixth international conference on computers in agriculture* (pp. 147–153). St. Joseph: American Society of Agricultural Engineers.

Aerts, J. M., Buyse, J., Decuypere, E., & Berckmans, D. (2003). Order identification of the dynamic heat production response of broiler chickens to step changes in temperature and light intensity. *Transactions of the ASAE, 46*(2), 467–473.

Allison, J. M., White, J. M., Worley, J. W., & Kay, F. W. (1991). Algorithms for microcomputer control of the environment of a production broiler house. *Transactions of the ASAE, 34*(1), 313–320.

Baras, E. & Lagardère, J.-P. (1995, June). Fish telemetry in aquaculture: Review and perspectives. *Aquaculture International, 3*(2), 77–102. ISSN 0967-6120.

Barnett, J. L., & Hemsworth, P. H. (1990). The validity of physiological and behavioral measures of animal welfare. *Applied Animal Behavior Science, 25*, 177–187.

Berckmans, D., & Goedseels, V. (1986). Development of new control techniques for the ventilation and heating of livestock buildings. *Journal of Agricultural Engineering Research, 33*, 1–12.

Brown-Brandl, T. M., Beck, M. M., Schulte, D. D., Parkhurst, A. M., & DeShazer, J. A. (1997). Physiological responses of turkeys to temperature and humidity change with age. *Journal of Thermal Biology, 22*(1):43–52.

Brown-Brandl, T. M., Yanagi, Jr., T., Xin, H., Gates, R. S., Bucklin, R. A., & Ross, G. S. (2003, September). A new telemetry system for measuring core body temperature in livestock and poultry. *Applied Engineering in Agriculture, 19*(5), 583–589. ISSN 0883-8542.

Czarick, M., & Lacy, M. P. (1994). *Environmental controllers. Poultry Housing Tips, A Cooperative Extension Report*. The University of Georgia Extension Service.

De Shazer, J. A., & Randall, J. M. (1988). Electronic stockmanship- present and future. In *Livestock Environment III, Proceedings of the 3rd Livestock Environmental Symposium*, 462–468, Toronto, Canada, 25–27 April. St. Joseph: ASAE.

Silva, A. C. S de, Arce, A. I. C., Souto, S., & Costa, E. J. X. (2005, November). A wireless floating base sensor network for physiological responses of livestock. *Computers and Electronics in Agriculture, 49*(2), 246–254. ISSN 0168-1699.

Flood, C. A., Jr., Trumbull, R. D., Koon, J. L., & Brewer, R. N. (1991). Partitioned ventilation control for broilers. *Transactions of the ASAE, 34*(6), 2541–2549.

Geers, R., Berckmans, D., & Huybrechts, W. (1984). Mortality, feed efficiency and carcass value of growing pigs in relation to environmental engineering and control: A case study of Belgian control farming. *Livestock Production Science, 11*, 235–241.

Goedseels, R., Geers, B., Truyen, P., Wouters, K., Goossens, H. V., & Janssens, S. (1992). A data-acquisition system for electronic identification, monitoring, and control of group-housed pigs. *Journal of Agricultural Engineering Research, 52*, 25–33.

Green, A. R. & Xin, H. (2009, December). Effects of stocking density and group size on thermoregulatory responses of laying hens under heat-challenging conditions. *Transactions of the ASABE, 52*(6), 2033–2038. ISSN 0001-2351.

Hahn, G. L. (1997). Dynamic responses of cattle to thermal heat loads. *Journal of Animal Science, 77* (suppl_2):10

Hamrita, T. K., & Conway, R. H. (2017a). First order dynamics approaching of broiler chicken deep body temperature response to step changes in ambient temperature. *International Journal of Agricultural and Biological Engineering, 10*, 13–21.

Hamrita, T. K., & Conway, R. H. (2017b, March). Effect of air velocity on deep body temperature and weight gain in the broiler chicken. *The Journal of Applied Poultry Research, 26*(1), 111–121.

Hamrita, T. K., & Hoffacker, E. C. (2008). Closed-loop control of poultry deep body temperature using variable air velocity: A feasibility study. *Transactions of the ASABE, 51*(2), 1–12.

Hamrita, T. K., & Mitchell, B. (1999). Poultry housing environment control: A summary of where we are and where we want to go. *Transactions of the ASAE, 42*(2), 479–483. (Hamrita, and Mitchell, 1999).

Hamrita, T. K., & Paulishen, M. (2011). Advances in management of poultry production using biotelemetry. In O. Krejcar (Ed.), *Modern telemetry* (pp. 165–182). InTech. INVITED. (Hamrita, and Paulishen, 2011).

Hamrita, T. K., Hamrita, S. K., Van Wicklen, G., Czarick, M., & Lacy, M. (1997). *Use of biotelemetry in measurement of animal responses to environmental stressors* (ASAE Paper No. 97-4008). St. Joseph: ASAE.

Hamrita, T. K., Wicklen, G. V., Czarick, M., & Lacy, M. (1998). Monitoring poultry deep body temperature using biotelemetry. *Journal of Applied Engineering in Agriculture, 14*(3), 11–15, , 227–231.

Havenstein, G. B., Scheideler, S. E., Ferket, P. R., Qureshi, M. A., Christensen, V., & Donaldson, W. E. (1991). A comparison of the 1957 Athens/Canadian random bred control strain with the 1991 Arbor Acres broiler when fed diets typical of those fed in 1957 and 1991. In *Proceedings of the North Carolina State Poultry Supervisors' Short Course*. April 1, 1992.

Korthals, R. L., Hahn, G. L., Nienaber, J. A., McDonald, T. P., & Eigenberg, R. A. (1992). *Experiences with transponders for monitoring bioenergetic responses* (ASAE Paper No. 92-3010). St. Joseph: ASAE.

Lacey, B., Hamrita, T. K., Lacy, M. P., & Van Wicklen, G. L. (2000a, June). Assessment of poultry deep body temperature responses to ambient temperature and relative humidity using an online telemetry system. *Transactions of the ASAE, 43*(3) 717–721. ISSN 0001-2351.

Lacey, B., Hamrita, T. K., Lacy, M. P., Van Wicklen, G. L., & Czarick, M. (2000b, Spring). Monitoring deep body temperature responses of broilers using biotelemetry. *Journal of Applied Poultry Research, 9*(1), 6–12. ISSN 1056-6171.

Lacey, B., Hamrita, T. K., & McClendon, R. W. (2000c, May). Feasibility of using neural networks for real-time prediction of poultry deep body temperature responses to stressful changes in ambient temperature. *Applied Engineering in Agriculture, 16*(3), 303–308. ISSN 0883-8542.

Lamade, R. M. (1984). *Computers in broiler growing operation* (ASAE Paper No. 84-4028). St. Joseph: ASAE.

Mader, T. L., Holt, S. M., Hahn, G. L., Davis, M. S., & Spiers, D. E. (2002). Feeding strategies for managing heat load in feedlot cattle. *Journal of Animal Science, 80*, 2373–2382.

May, J. D., & Lott, B. D. (1992). Feed and water consumption patterns of broilers at high environmental temperatures. *Journal of Poultry Science, 71*(2), 331–336.

Mitchell, B. W. (1981). *Effect of handling and temperature stress on the heartrate, EKG, and body temperature of chickens* (ASAE Paper No. 81-4543). St. Joseph: ASAE.

Mitchell, B. W. (1984). Interfacing single-board microcomputer controls to conventional controls for an environmental control system. *Transactions of the ASAE, 27*(5), 1590–1594.

Mitchell, B. W. (1986). Microcomputer-based environmental control system for a disease- free poultry house. *Transactions of the ASAE, 29*(4), 1136–1140.

Mitchell, B. W. (1993). Process control system for poultry house environment. *Transactions of the ASAE, 36*(6), 1881–1886.

Moriarty P. (1993,Summer). Foodborne pathogens-new controls from farm to table. *Food News for Consumers*, (Supplement): 4.

Morton, D. B., Hawkins, P., Bevan, R., Heath, K., Kirkwood, J., Pearce, P., Scott, L., Whelan, G., & Webb, A. (2003, October). Refinements in telemetry procedures. *Laboratory Animals, 37* (4), 261–300. ISSN 0023-6772.

Payne, C. G. (1966). Practical aspects of environmental temperature for laying hens. *World's Poultry Science Journal, 22*(2), 126–139.

Schurmas, B., Berckmans, D., Decuypere, E., Buyse, J., & Aerts, J. M. (1996). Description of an automated open-circuit indirect multi-calorimetry system suitable for dynamic measurements. *Journal of Applied Physiology*. (In Review).

Shlomo, Y., Goldfeld, S., Plavnik, I., & Hurwitz, S. (1995). Physiological responses of chickens and turkeys to relative humidity during exposure to high ambient temperature. *Journal of Thermal Biology, 20*(3), 245–253.

Simmons, J. D., & Lott, B.D. (1993). Automatic fan control to reduce fan run time during warm weather ventilation. *Journal of Applied Poultry Research, 2*(4):314–323.

Steinbach, J. D. (1971). Relative effects of temperature and humidity on thermal comfort in swine. *Nigerian Agricultural Journal, 8*, 132–134.

Tao, X., & Xin, H. (2003a, April). Acute synergistic effects of air temperature, humidity, and velocity on homeostasis of market-size broilers. *Transactions of the ASAE, 46*(2), 491–497. ISSN 0001-2351.

Tao, X., & Xin, H. (2003b). Temperature-humidity-velocity index for market-size broilers. In *Agricultural and biosystems engineering conference proceedings and presentations. Paper 197.*

Timmons, M. B., & Gates, R. S. (1987). Relative humidity as a ventilation control parameter in broiler housing. *Transactions of the ASAE, 30*(4), 1111–1115.

Timmons, M. B., Gates, R. S., Bottcher, R. W., Carter, T. A., Brake, J., & Wineland, M. J. (1995). Simulation analysis of a new temperature control method for poultry housing. *Journal of Agricultural Engineering Research, 62*(4), 237–245.

Worley, J. W., & Allison, J. M. (1984). *Microprocessor control of poultry house environment* (ASAE Paper No. 84-3025). St. Joseph: ASAE.

Yanagi, Jr., T., Xin, H., & Gates, R. S. (2002, March). A research facility for studying poultry responses to heat stress and its relief. *Applied Engineering in Agriculture, 18*(2), 255–260. ISSN 0883-8542.

Yang, H. H., Bae, Y. H., & Min, W. (2007). Implantable wireless sensor network to monitor the deep body temperature of broilers. In *SERA 2007. 5th ACIS International Conference on Software Engineering Research, Management & Applications* (pp. 513–517). Busan: IEEE.

Zhang, G. (1993). A PC-based multicompartment climatic control system for agricultural buildings. *Computers and Electronics in Agriculture, 8*, 211–225.

Dr. Takoi Khemais Hamrita is a Professor of Electrical Engineering at the University of Georgia, where she has spearheaded the development of two ABET-accredited degree programs, one in electrical and the other in computer systems engineering. These efforts have recently culminated into a new UGA school of electrical and computer engineering for which she has served as inaugural chair. Dr. Hamrita has been at UGA for 25 years, where she was the first woman faculty member to be hired into the Department of Biological and Agricultural Engineering as an Assistant Professor (she remained the only female Professor in her department for almost 15 years). Dr. Hamrita's main research focus is on precision agriculture. She has worked in many areas of precision agriculture including yield monitoring, smart poultry environmental control, biotelemetry, smart sensing, and harvest and post-harvest technology. She holds a patent in yield monitoring, has published over 50 articles and book chapters, and has given over 50 conference presentations around the world on related research. Dr. Hamrita has served in many leadership roles within professional societies including:

- Chair, Vice Chair, and Secretary of the IEEE-IAS (Industry Applications Society) – Industrial Automation and Control Committee
- Associate Editor for the Industry Applications Society journals
- IEEE-IAS Conference technical program chair
  Southeast coordinator and IEEE USA liaison for Women in Engineering (WIE).
- Chair, Vice Chair, and Secretary of the ASAE (American Society of Agricultural Engineers)-IET (Instrumentation and Control Committee)
- Comparative and International Education Society (CIES) session organizer at the CIES Annual International Conference
- American Society of Engineering Education (ASEE) session organizer at the ASEE Annual International Conference.

Dr. Hamrita is a fierce advocate for women in STEM and is the founder and chair of the Global Women in STEM Leadership Summit. The program is an ongoing movement that aims to educate, inspire, empower and elevate women and girls in scientific and technological fields to help eliminate internal and external barriers they face. The women come from all career paths and stages from high school to the C-Suite. Our goal is to build capacity, nurture talent, create community, and empower women and girls in STEM to reach their full potential and thrive in male dominated fields. The summit convenes some of the most successful and influential leaders and founders from industry, academia, nonprofit and government. Dr. Hamrita has built a decade-long partnership between UGA and the Tunisian Ministry of Higher Education, which has had profound impact both on UGA and Tunisia and has become an innovative model for education and development around the world. Some of the most notable impacts of this partnership is the launching of a national virtual university in Tunisia that is currently delivering, online, a sizeable portion of higher education curricula across disciplines. The program has earned her numerous prestigious awards such as the Tunisian National Medal in Science and Education, the Andrew Heiskell Award for Innovation in International Partnerships, and the Tunisian Community Center's Ibn Khaldoun Award for Excellence in Public Service.

Taylor Ogle is an undergraduate student who graduated in May, 2020 with a degree in agricultural engineering from the University of Georgia. Her area of emphasis is natural resource management. She is involved with the Society of Women Engineers (SWE), the College of Engineering Student Ambassadors, and the Baptist Collegiate Ministries (BCM).

Amanda Yi is an honors student in her last semester at the University of Georgia. She is earning a bachelor's degree in electrical engineering. She mentors elementary and middle school students in the Athens community through the program "Whatever It Takes". After graduation, Amanda will work as an electrical engineer in the aerospace manufacturing industry.

Chapter 9
Advancement in Livestock Farming through Emerging New Technologies

Jarissa Maselyne

Contents

9.1 Background and the Start of My Career

I am an electromechanical engineer by education. Why? Because I could. Becoming a veterinarian or a biologist was discouraged by my environment because allegedly, there was no job security in those areas. I had good results in science and maths so I chose something in that direction. Also at school, some teachers more or less subtly pushed me into choosing engineering. So it was that which I chose, by process of elimination. Bioscience engineering was very plant, lab, and food oriented, which didn't appeal to me, so I went for civil engineering.

After the first couple of general years in civil engineering major, I chose electromechanical engineering as my main direction of study, again because this allowed the largest variety of possibilities. Then came the choice of a major in the master's program, and it occurred to me that none of the directions (maritime engineering, mechanical energy engineering, electrical power engineering, etc.) actually interested me. In the end, I went for control engineering and automation, which seemed like something I could enjoy. But the more I learned throughout those 5 study years about the professional careers you could have with the degree, the more I wondered

J. Maselyne (✉)
ILVO (Flanders Research Institute for Agriculture, Fisheries and Food), Technology and Food Science Unit – Agricultural Engineering Research Area, Merelbeke, Belgium
e-mail: jarissa.maselyne@ilvo.vlaanderen.be

© Springer Nature Switzerland AG 2021
T. K. Hamrita (ed.), *Women in Precision Agriculture*, Women in Engineering and Science, https://doi.org/10.1007/978-3-030-49244-1_9

what I was actually doing in this study! As you can imagine, at that age, with my life choices ahead of me, I questioned my future and tried to find back the intrinsic interests that I seemed to have lost.

And what interests me the most is animals. I had several pets as a child and was happiest with them around me. As a 10-year-old, I studied the chickens in the yard through the window and wrote down their behaviors. Then the only thing I ever defied my parents for was to be able to go horse-riding. I was known for having a stock of knowledge about wild and domestic animals that I just had no trouble remembering. So after some job applications in several fields, and an extensive search for jobs related to animals that didn't fit my degree, I stumbled across an open PhD position at the KU Leuven in cooperation with ILVO (the Flanders Research Institute for agriculture, fisheries, and food). The title of the position was "Researcher in innovative technologies for pig husbandry." This sounded like music to my ears, combining my knowledge of technology, data analysis, and process control with an aim to optimize health and welfare of pigs at the individual level. So I thought, this is it! Only one issue, the application mentioned that you needed a degree in Veterinary Sciences or Bioscience engineering, neither of which I had. So I called the contact persons and asked if I was eligible. They told me to file my application anyway.

To my surprise, I was invited for a job interview, and the days before, I studied everything I could find about pig farming on the Internet and read lengthy reports about sow housing systems. I had never been in a pig barn and knew nothing about it. At the job interview, I was competing with people that had the right degree, and I got questioned by a very large group of jury members sitting at a long table. There was one detailed question about pig farming (I remember it very well) from Annelies Van Nuffel, who later on became my closest supervisor. She asked: Do you know what a 3-week batching system is? Somewhere in those lengthy reports I had come across it, so half guessing, I replied that it is a kind of management system to group the actions you need to do on your sows. I got sent out of the room, and less than 15 min later (I wasn't even on the bus yet), I got called that they wanted me for the job!

And that's how I started at ILVO, on a PhD research together with the KU Leuven and Prof. Wouter Saeys. I got a scholarship for the PhD by writing an application and defending the subject for a funding authority (former IWT Flanders), and my main working space became ILVO, which was closer to my home than Leuven and the place where the experimental facilities were. Annelies became my ILVO supervisor, and I got to visit my very first pig barn together with a swine vet colleague, Liesbet Pluym, and spent an intro day working together with the farm caretakers to learn the daily practices at a farm. Already in my first working weeks, we had a meeting of the European project PigWise (ICT-Agri Era-net, 2011–2013) in Germany. Additionally, in the early months, I was tasked with arranging the renovation of a department of the pig farm for my experiments. So working with the team of ILVO technicians, I designed the barn, contacted suppliers, calculated the size of the ventilation system and the windows, decided on the type of feed delivery system, and installed a new silo. Quite a variation in activities, so it was never boring!

9.2 Automated Monitoring of Feeding and Drinking Patterns in Growing-Finishing Pigs

The topic of my PhD originated from the PigWise project that was devised from the ideas of the coordinators Engel Hessel (University of Göttingen at the time) and Kristof Mertens (one of my KU Leuven supervisors at the time). The idea was to use RFID tags (radio frequency identification) in the ears of the pigs to have an electronic ear tag with a unique ID. Then, antennas installed around the feeders could monitor each pig's presence at the feeder and thus derive feeding visits and durations from it. Pigs, like humans, could be expected to change their feeding behavior when they are under stress, due to disease or welfare issues. In addition, feeding is vital to the pig's growth and performance. So by monitoring their behavior from day to day, abnormal changes could be detected and relayed to the farmer as an indication of a pig becoming ill or having a problem.

The system is summarized in Fig. 9.1, where RFID is used for data collection, and then algorithms are applied to the data to construct alerts and finally give these alerts as feedback to the farmer so he or she can take the proper actions for the pigs. In total, this forms a warning system for performance, health, and welfare problems in individual pigs as an extra pair of eyes and ears in the barn. What was and still is very innovative about this is that individual pigs are automatically monitored. You must know that in regular pig farming, the management is mainly at a group level. Pig farmers in Belgium have around 2000 pigs on average for fattening (Statistics Flanders, 2019), and these are grouped in pens of 10–20 animals or even up to 200–300 pigs in the same group. Problems in the pig barn are currently spotted through a visual inspection round by the farmers once or twice a day. Working time

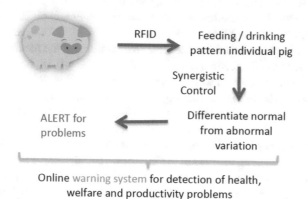

Fig. 9.1 Concept of the warning system for detection of health, welfare, and productivity problems: Feeding and drinking patterns of individual pigs are measured with RFID technology, and Synergistic Control is applied to the data to differentiate normal variation from abnormal variation and alert the farmer of problems (*RFID* = Radio Frequency Identification). (Reproduced from Maselyne 2016)

investment for this livestock check has been measured to be between 3.6 and 6 s per pig per day (Heitkämper et al. 2011; van den Heuvel et al. 2004). It is at the moment practically impossible for the farmer to identify and recall an individual animal and its disease history without the help of (digital) tools. The limited time available, large groups of animals, prey animal behavior (hiding disease from predators), observer bias, and the difficulty to detect illness, lameness, and welfare issues generally indicate the need for a supplementation with automated monitoring.

Being an engineer, I decided to validate each subpart of the system before being able to draw general conclusions. So I validated the RFID system both "off-line", so without pigs but with a whole grid of tags at fixed positions in various orientations and distances from the antenna (Maselyne et al. 2014b), and "online" via video scorings of feeding pigs as reference data (Maselyne et al. 2014a). Then I researched and validated the best method to construct feeding visits from the RFID registrations, again compared to video scorings (Maselyne et al. 2016). For these very time-consuming scorings, I was lucky to have the help of Petra Briene, a student at the time. Finally, I reviewed the existing literature about the process of forming meals (clusters of feeding visits close in time that end, after which a pig or any other animal is considered to be satiated) (Maselyne et al. 2015c). The numerous methods that I found were often not based on an underlying motivation, and the sporadic evidence (which looks to be confirmed by my data) that pigs in high competitive environments do not always show clear meals made me leave that topic at the theoretical level and focus further only on visits and raw registrations.

Several PhD years had flown by during which I also had four fattening rounds of pigs that needed follow-up. Monitoring the pigs' health, solving technical issues, and checking the data was a continuous job in between the validation studies and papers. Sometimes a hassle, having to change clothes, shower, and go to the barn frequently, and sometimes a nice break during the day. At the same time, the RFID drinking system was designed (Fig. 9.2), and Ines Adriaens, a master's student, validated the system, which resulted in a joint paper (Maselyne et al. 2015b) and a book chapter (Maselyne et al. 2015a).

As conclusions, we found that the RFID registrations of the feeding system were well correlated with the feeding duration of the pigs ($R^2 = 0.88$) but only when four pigs of the 20 observed pigs were removed (with these pigs, $R^2 = 0.53$) (Fig. 9.3a) (Maselyne et al. 2014a). These four pigs' behavior impaired the correlation because they were either frequently lying down while feeding (which led to less registrations than expected) or they were standing next to the trough without accessing the feed (which led to more registrations than expected) (Fig. 9.3b, c) (Maselyne et al. 2014a). When analyzing the pigs in the group, we could observe that their behavior was a habit and sometimes also related to their characters and their ranking in the group (low-ranked pigs needing to wait for access to the feeder, higher-ranked pigs guarding access to the feeder). This was good news because if we compare each pig to its own normal behavior, we can compensate for the inferior correlation some specific pigs have between registrations and actual feeding.

However, it was also important to analyze in more detail the technical reasons why this happened. Therefore, we conducted very extensive range measurements

Fig. 9.2 One pen in the experimental barn at ILVO, with the RFID feeding and drinking system installed to monitor each individual pig's feeding and drinking patterns (*RFID* = Radio Frequency Identification) ·

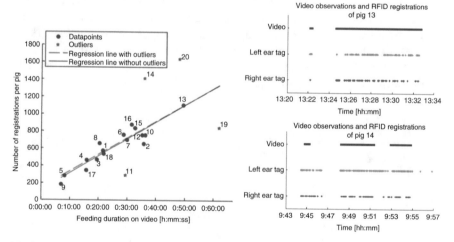

Fig. 9.3 (**a**) Linear regression of the number of RFID registrations per pig versus the feeding duration on video for each focal pig with all pigs included (y = 143.14 + 0.32 ×, R² = 0.53) or with four outliers removed (y = 117.71 + 0.33 ×, R² = 0.88). (**b**) Video observations and RFID registrations of left tag and right tag of pig 13, a pig with an average ratio number of registrations versus feeding duration, and (**c**) of pig 14, an outlier with a high ratio number of registrations versus feeding duration (*RFID* = Radio Frequency Identification) (Reproduced from Maselyne et al. 2014a)

around the antennas placed at the feeder. We found that tag position and orientation had a significant effect on whether or not a tag is detected by the antenna (Maselyne et al. 2014b). Theoretically, this was consistent with how this type of RFID system worked, and now we were able to quantify it for our system. Specificity was very high, and the range only covered the feeding trough, but sensitivity was not good when the tags were too close to the ground (Maselyne et al. 2014b). This explains why lying pigs were not registered well, and this might be a risk for young pigs when they arrive at the barn the first weeks or months. Since then, we adjusted the height of the antenna to the average height of the pigs. We also learned from these tests that the variable gaps between registrations during the same feeding visit that we observed (Fig. 9.3b, c) were likely caused by the constant movement of the pigs and related change in orientation of the tags in their ear (Maselyne et al. 2014b).

To replicate feeding visits, we needed to combine the RFID registrations into visits. Several methods were tried and compared with video observations as reference data (Maselyne et al. 2016). We determined that RFID registrations could be bundled using a single-bout criterion of 10 s with two tags on each pig or 20 s if each pig had just one tag (Maselyne et al. 2016). Registrations with less than or equal to the bout criterion between them are considered part of the same feeding visit. Forming visits led to an improved relationship between RFID and observed feeding (Fig. 9.4) ($R^2 = 0.86$ instead of $R^2 = 0.73$ across two experiments), but also had the risk of having a variable performance among pigs, although the method was repeatable across pigs and days and reproducible across experiments (Maselyne et al. 2016). Therefore, pretreating any data must always be done with care. From the analysis and video observations, some other interesting conclusions could be made. First, performance of the system was better with two tags than with one, and so for further experiments, we equipped pigs with a tag in each ear despite the

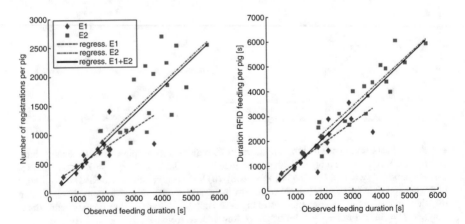

Fig. 9.4 Linear regression of the observed feeding duration versus **a** the number of RFID registrations per pig in two experiments (*E1* + *E2*) (y = −52 + 0.5x, $R^2 = 0.73$) and **b** the duration of RFID-based feeding per pig (*E1* + *E2*) (y = −38 + 1.1x, $R^2 = 0.86$) (Note that in E1, only one out of two feeders in the pen was observed; data shown are for two tags per pig) (*RFID* = Radio Frequency Identification). (Reproduced from Maselyne et al. 2016)

increase in cost. Also, feeding visits were very short (33–42 s on average) and close to each other, showing a different feeding behavior to other feeding systems with feeding stations, for example (Maselyne et al. 2016). Indeed, with this system, it now became possible to record feeding behavior under commercial farm circumstances at large scale.

For the drinking system, no range measurements were needed as the antenna and reader system were the same. Antennas were placed vertically around the nipple drinker instead of horizontally above the feeder. To avoid registering pigs in the neighborhood, we added plastic barriers (triangles) on either side of the nipple (Fig. 9.2). Average drinking visits were also short (17–28 s on average), and the same bout criterion was used (Maselyne et al. 2015b). A minimum (3 s) and maximum (3 min) duration criterion were added to remove short passages along the antenna and playing or lying bouts around or under the antenna. Again having two tags per pig had a benefit in that it helped to further remove false bouts where only one ear was in range of the antenna (for a drinking pig, you would expect registrations of both ears) (Maselyne et al. 2015b). Filtered number of registrations and constructed drinking visit's duration were correlated with the observed drinking duration ($R^2 = 0.90$ and $R^2 = 0.88$) and the registered water usage for every bout ($R^2 = 0.75$ and $R^2 = 0.71$) (Maselyne et al. 2015b).

All of this taught me a great deal about the system and the imperfections it has. As I learned, a thorough step-by-step validation is a crucial and too frequently forgotten part of sensor and system development as it revealed what to look out for and how reliable the data was. In data models, almost always you have the phenomenon "rubbish in, rubbish out", so blindly analyzing any data you get is not the right way to do it. With a good understanding of the data, it was now time to work on the end goal: the warning system. I had written a lot of pieces of code already, did some preliminary analysis, and tried some warning systems. So my advisor pushed me to design and select the most promising warning systems in a couple of weeks' time and to start up a final measurement round for validation in the last phase of my PhD.

9.3 Toward a Warning System for Performance, Health, and Welfare Problems in Individual Pigs

When designing a warning system, we first needed to know what it should actually warn for. We considered health, welfare, and productivity problems, but there is a range of these problems as they all come in different variations, degrees, and combinations and with different effects on the pig itself. Initially, we planned to focus our efforts on the treatments and the notes of the barn caretakers. However, in the previous fattening rounds, I used to go into the barn a couple of times a week as well to check on the pigs, and during weighing (every couple of weeks). This gave me the opportunity to make more detailed observations of things like lameness, lesions, and body condition score. Only the severe cases were usually treated, and different people might spot different things in the same group of pigs. If you could spend more time per pig than an average farmer, it's possible to become more sensitive to mild problems or detect them earlier, and with the relatively small group of pigs we had in the barn (140–236 pigs at any time), it becomes possible to compare each pig to their own normal individual activity levels. So do we not want our system to be better than the general visual inspection? How can we reduce the subjectivity in the observations?

Together with some colleagues from ILVO and the swine health department of Veterinary Sciences at the University of Ghent, we made a scoring sheet that covered four pages of visual symptoms that a vet looks out for when checking the overall status of a pig. From the discussions and our experiences with individual pig observations, we formed a list of criteria for when a warning should be given to the farmer and when not. We decided to check every pig every weekday inside the pens and in addition check the pigs that would give an alert for one of the warning systems in more detail afterwards. We would weigh and score the pigs every 2 weeks, have a vet check every 2 weeks, and send every deceased or euthanized pig for an autopsy. Also at slaughter, several extra things were scored like lung and liver lesions and the presence of internal abscesses. For a period of 4 months, Petra – who volunteered to help me – and I were in the barn every weekday for close to half a day, one of us writing down all the notes and one of us checking the 140 pigs in the pens, making them stand up and walk to check for lameness, score, and analyze them from top to bottom, taking their temperature when we suspected fever and making sure we were not eaten by the ones that wanted to play.

Going forward, we experienced the need to not only have a category of pigs that definitely required an alarm (red status) and ones that definitely did not require an alarm (green status), but there were also a number of mild problems, doubt cases, that would be helpful to give an alarm, but not a crucial or certain one (orange status). Therefore, we created an in-between reference category. Even then, it was difficult to cover all possible cases. A diseased pig could go up and down in how it felt from day to day, and even if we tried to see them all every day, we didn't know how they felt the rest of the day. Also, two pigs with the same symptoms could react very differently, for example, two pigs that are equally lame where one would be still active and feeding normally and the other would be depressed and sore. Therefore, activity levels of the pigs were something we needed to take into account. Even then, we noticed different "normal" levels in the pigs, so it was hard to keep track of this for 140 individuals (let alone if you would have a farm of 3000 in practice). Also, trying to score a pig when it had just been sleeping versus when it was in play mode added another challenge. Adding to that, we had a pig that did not have any visual symptoms, but we got alerts for it almost every day, and the only thing we could say was that it looked a bit "off". But according to our list of symptoms and categories, it was still in the green status. It had droopy ears which made it stand out for the caretakers and vets at several occasions, because this could be a sign of issues, but nothing else was found. However, at slaughter, we found it did have internal abscesses! On the other hand, the droopy ears could have caused it to be registered less well at the feeder during feeding and therefore gave more alerts. So we cannot be sure whether these were true or false alerts.

After the fattening round, all the observations and other information were gathered, and for every pig and day of the period, a status of red, orange, or green was given. This was then compared on a one-to-one basis with the alerts a certain warning system would give. The designed warning systems were based on the principle of Synergistic Control and trained on historical data. Synergistic Control is basically a combination of modeling the normal trend in the data (Engineering Process Control) and applying control charts to detect the abnormal variation in the data (Statistical Process Control). This allows to treat each pig as an individual and to only detect days that are abnormal compared to the pig's own history but taking into account that a pig changes a lot during its growth period. Several sensitizing rules could be added, for example, to avoid missing pigs that start the round with some issues already and thus a very low level of feeding. We compared these individual, time-varying control limits with fixed limits that are the same for all pigs and all measurement days, and both methods were applied to two monitoring variables each: the daily number of registrations and the daily average gap between feeding visits (Maselyne et al. 2018).

The best performance was for the Synergistic Control method on the number of registrations: 58.0% sensitivity, 98.7% specificity, 96.7% accuracy, and 71.1% precision (Maselyne et al. 2018). Severe problems (red status) that lasted more than 1 day were detected in 64.4% of cases and this within 1.4 days from the start on average (84.5% at the first day) (Maselyne et al. 2018). If we would exclude the two pigs with long extended periods of false alerts of a doubtful status (47 and 17 alerts, resp.), the precision would be 79.4%. Both of these pigs had droopy ears (one was already described above), but both of them also had serious chronic health issues for an uncertain period of time (Maselyne et al. 2018). Looking at missed problems, a frequent observation was that these pigs gradually developed a problem quite early in the fattening period, and the model would follow this gradual decrease, or the control limits would become very wide due to the large variation at the start, thus leading to the problem period not being detected (thus regarded as "normal"). A solution to this problem could be to initialize the model and control limits based on historical data of the behavior of an average healthy pig and gradually bring in the specific individual pigs' behavior to tune the warning system to every new pig individually. This prevents that strange behavior at the beginning of the round is considered normal for a certain pig. Some post hoc analysis to try out this strategy revealed that the sensitivity could increase with 8 percentage points, so to around 66% (Maselyne 2016).

Looking critically at this performance evaluation, we identified a number of issues. First of all, the orange days are very difficult to handle as they are defined as "okay if they give an alert, but not really needed": in essence, this means the warning system cannot be wrong on these days. So the more orange days you consider, the higher the performance numbers will be, artificially. Second, an alert that is given for a pig without visual symptoms is always considered false, but it can be a subclinical disease or internal issues (like the example of the abscesses above). A pig that is still showing visual symptoms, but has started eating again because it is recovering, will be recorded as a missed problem just because there is a lag in how fast the clinical symptoms disappear.

The performance of the system would need to improve in terms of sensitivity and precision, but we could never expect every visual symptom to have an effect on the feeding and this effect to be uniform for all pigs. Additionally, a number of things that happen with the pigs cannot be seen by us checking them once a day (or at all; will your colleagues see it when your back is sore or you have a headache and you don't say anything?). Health, welfare, and productivity problems are such a broad spectrum of complicated mechanisms, and using one variable to predict them all is an overly ambitious goal. Is it then a worthless system? No, quite the opposite. First of all, visual inspection is not a good reference on its own either; thus, the feeding data and visual inspection should be used together in order to achieve the best results. What really illustrates this is that each time the vet did a checkup of the barn, she asked me when she spotted a pig with issues, "Is that pig still eating?" And I would reply, "I cannot tell you that because we are validating the system based on your assessments." But she asked so many times because it was such a crucial information for her to decide if the pig needed treatment or not, information she was otherwise always lacking and she knew I had. In addition, the data show some really interesting things: like the feeding behavior going up and down in parallel with the fever measured in a pig. Also, it seems to be possible to spot in the feeding behavior, whether or not a treatment was working, days before the pig was visually okay again.

9.4   A Moment of Reflection

December 6, 2015, I had my internal PhD defense on my birthday, and on February 4, 2016, I received the PhD degree at the public defense. Do I still look back at my master's degree as a wrong choice? No, it gave me opportunities I might not have gotten otherwise, and it allowed me to also tackle the PhD topic from a technical perspective, but still combining it with livestock knowledge through the practical experiences in the barn. I don't actively use much of the knowledge I learned in the courses anymore, but I believe that's the case for a lot of people. And I suppose it does help to tackle problems, find a solution, and further understand some concepts.

Looking back at the achievements during my PhD, I think most people that finalized a PhD have the same feeling: I wish I could have done more because there are so many more pieces of the puzzle needed. Could there be more gain in the monitoring of other variables than the ones we tested? Or maybe using multiple variables at the same time, and what about the drinking behavior? What about other types of algorithms or more combinations of using historical data and the individual pigs' data? How would the system perform on another farm? What is really the added value for the farmer, and how does it affect his/her time, cost, or ease of working? What is the cost-benefit ratio, how much could the prototype decrease in price when marketed, how much costs need to be calculated for software, and how much can you gain from better monitoring of the pigs? How can a farmer handle the warnings and find the specific pig that needs attention in the pens? Are the warnings not too sensitive at the moment (because our observations as researchers are very sensitive compared to the caretakers), and will he/she not be overwhelmed by alerts? So all of these remain open questions and work to do, but the first steps have been made.

There's currently some debate about "early" warning systems in the topic of precision livestock farming. Over the years, we deliberately avoided the word early, because what does early mean? We all agree we shouldn't be late, but early is linked with detection before clinical symptoms are visible. However, what should a farmer do when he/she gets a warning for a pig before the symptoms are visible? How can he/she determine that the warning is true? How can he/she find the problem and give the right treatment? I think this is a very difficult matter that needs to be further explored. In addition, at the moment I believe veterinary treatment for pigs is mostly focused on group-level issues and not yet on individual pigs. In addition, the treatments are quite similar for a large variety of problems. So also, on the level of treatment, a lot of work is yet to be done.

When presenting the work, I often get the question whether it is possible to detect what problem the pig has. At the moment, this is not yet possible. First of all, because it is rare that we see the same issue twice (with same severity and developing in the same way), and often a pig develops several problems at the same time or in a short time span. Secondly, the algorithms already have difficulty detecting if there is a problem in the highly variable data. But, as mentioned, the first steps have been made and there is potential. Current data as well as future multivariable data is valuable information for the farmer and the veterinarian to determine the type and severity of the problem. A very abrupt change in feeding will be more likely to occur due to an accident or a very fast developing disease (like cardiovascular failures or acute pneumonia) compared to a slowly decreasing feeding behavior. If a pig starts to eat less and drink more, it could have, for example, heat stress or diarrhea problems. So it does give an indication, and the more data would be added, such as environmental data, pig weight, and pen averages, the more information can be extracted.

Looking at the potential of such a warning system based on RFID, the benefits can be present at many different levels: first of all, increased health, welfare, and productivity of the individual pigs. The alerts can lead to improved objective decision-making by the farmer, a more efficient use of labor and time, and economic returns. The better condition of the pigs, and more control of the farm, can also give the farmer more satisfaction in his/her work. Working at the individual level and based on detailed measurements is definitely also a way to improve the sound use of treatment and antibiotics and thus also reduce antimicrobial resistance. Because of the RFID tags, traceability and data integration become possible at a much more advanced level than is the case now. Finally, the tool can show to consumers the good follow-up and standard of life of the pigs and thus increase consumer trust and satisfaction.

During my PhD, the topic was considered very innovative and seemed a distant future for a lot of people from practice. At that time, a group from the USDA (United

States Department of Agriculture) had already developed a low-frequency RFID system for pigs and cattle. Not long after we started with the high-frequency RFID system in the PigWise project, the University of Hohenheim (Germany) started similar developments with an ultrahigh-frequency system. But those were the only people in the world that I knew were working on the same topic. Now, only 4 years later, RFID tags in general are finally becoming integrated in more and more practical fattening pig farms (which is already a first important step). Examples of this include 100 farms of the KDV pork chain in the Netherlands and farms and slaughterhouses of Tönnies Livestock in Germany. Before that, RFID tags were only used sporadically in sows for the purpose of giving their specific ration in an electronic sow feeder, but nothing more. In addition, research groups in France (IFIP, INRA), the Netherlands (WUR), and also Brazil, Australia, China, etc. are developing similar warning systems based on RFID-collected feeding and drinking behavior of pigs. I was also in contact with a producer's cooperative that wants to develop it and promote it for their farmers to gain a competitive advantage. At the end of my PhD, I got the opportunity to speak at MSD/Merck Animal Health conferences in Vienna, Miami, and Manila for an audience of their clients. So within a couple of years, the interest in these systems has really grown! I'm excited to see where the other developments go and what will be adopted and how.

The end of a PhD is also a turning point in a career, the end of a job of limited duration. Four months before, I got married and my husband was also due to finalize his PhD soon after. Many of our colleagues left for industry after their PhD, but we both received the opportunity to stay in research at ILVO and the University of Ghent, respectively, and we decided to take it. I stayed at the same office, same desk, and just switched supervisor in order to work more for the Agricultural Engineering Department as a whole (under the lead of scientific director Jürgen Vangeyte) instead of just the precision livestock farming group (with Annelies Van Nuffel, my former PhD supervisor, as a group leader).

9.5   Into Project Management of EU Flagship Innovation Projects

Early on during my PhD, I sporadically helped with project proposals at ILVO, such as writing or reading parts of the text or making budget calculations. At the end of 2015, I also ended up helping Jürgen in a H2020 proposal on Large-Scale Pilots for Internet of Things (IoT) in agriculture. We were asked to submit ideas for use cases, and I collected these from colleagues. After selection of the short list, two use cases of ILVO remained: one about tracking grazing cows and one about the follow-up of my PhD, installing, testing, and demonstrating the warning system at practical farms. For the latter, we were asked to join several pig-related ideas, like one about boar taint (an unpleasant odor that sometimes develops in the meat of uncastrated male pigs when it is heated), and we discussed with some other partners and companies to form one big use case about pig farm management. ILVO also became involved in the organization of the work packages and other initiatives in the proposal. Part of this work occurred during the last months of my PhD, during the first months of my new "postdoc" contract, and during a research stay in Finland.

I got the opportunity, through a short-term scientific mission funded by the COST action DairyCare, to go and work in Helsinki, together with Matti Pastell of Luke (the natural resources institute of Finland). First, Annelies, Matti, and I went to Foulum, the livestock department of Aarhus University, to meet with Lene Munksgaard and Vivi Thorup to discuss and collect accelerometer data of dairy cows and their hoof lesions recorded in a time span of almost 2 years. Then I stayed in Helsinki for 6 weeks to analyze this data with the help of Matti, and it resulted in a published short communication paper. The period in Finland was a nice break in between all the other work, where I could focus on one topic mostly and really go in depth into the literature and the data analysis again. I learned to program in R and learned about cow behavior and hoof lesions at the same time.

Together with Annelies, we tried for the second time to get national funding for a project about technology in pig farming (first time about electronic identification, second time about precision livestock farming), again without success. In the summer of 2016, we got the news that the IoT proposal, called IoF2020, was accepted! It started in January 2017, and we had two use cases, a task as chair for the meat trial and co-lead of the work package that monitors all the use cases. So from a research job, I went into a project management job with full-time writing projects and managing projects with loads of meetings, Skype calls, e-mails, reports and templates for reports, etc., and I officially became a direct supervisor of someone for the first time.

So my job content completely changed, and as for many postdocs probably, it got me thinking about what I liked the most. Actually, I missed my PhD time. I missed the time that I was able to go in depth into a topic, become an expert, do data analysis for days in a row, and really feel that I was making progress. Because although now I had a lot more things booked on my agenda, at the end of the day, I did not always have the feeling that I achieved something. Adding to that, continuous contacts with new people through mail, Skype calls, and events and a very much increased responsibility made it a struggle to get through. You must know that I am an introvert and my personal fulfillment depends partially on other people's perception. So each mail I had to send, I questioned how it would be perceived. Every Skype call I needed to chair, and I had plenty, was a process of conquering my own fears. Every event I had to present at was stress about what people thought of me. Because being responsible for a work package co-lead together with Jürgen, taking part in the Project Steering Group of this €30 million-funded H2020 project with over 70 partners, daily tasks of monitoring all dairy and meat use cases, and being trial chair and coordinating a use case, are all very ambitious things, so I really felt that I had to prove myself among all of the more confident and experienced men in the project.

In addition, IoF2020 was a very ambitious project, one of the flagship projects of the EU at the time and could thus also be a very demanding project – pushing all partners to their maximum in order to achieve the most. There was a lot of pressure on the use cases as well, to deliver the newest innovations on the market. So we needed to be at the top of the technological IoT innovations and at the same time develop business ideas, price settings, and distribution models while these business aspects were quite new to me.

9.6 Here Are the Buzzwords: IoT and HPC

Eventually, I adapted and learned, and perhaps I was not so bad at doing project management because quite quickly I got involved in different projects and project proposals for our Agricultural Engineering Department on different cutting-edge topics. So first of all, there was IoF2020. With a project that received one of the biggest funds in agriculture in the H2020 program, IoF2020 became a whole community of people with over 100 partners. IoF2020, or Internet of Food and Farm 2020, was aimed at a large-scale take-up of IoT in the European farming and food domain. IoT is basically creating smart webs of interconnected objects that are context-sensitive and can be identified, sensed, and (preferably) controlled remotely. In the agricultural domain, these objects can be tractors, plants, animals, and all the sensors that surround them. IoF2020 was built around 33 use cases, each of which aims at a certain development for a specific subsector.

Soon after the start of the project, Annelies, my former supervisor, left ILVO. My colleague Stephanie Van Weyenberg and I took over her responsibilities, and together we led the precision livestock farming group, where Stephanie took charge of the cow projects and I the pig projects. I was also responsible for the general follow-up of ILVO's tasks in IoF2020. We hired Evi Lippens to take over a large part of my project management duties in the project and Chari Vandenbussche to help with the pig use case, and soon after that, Petra also came to work for the project soon after which she started her own PhD research. I also took over the supervision of my colleague Shaojie Zhuang who was doing a PhD on precision livestock farming (PLF) and ammonia measurements from Annelies.

Another H2020 project we were involved in is CYBELE, which was aimed at fostering precision agriculture and PLF through the advancement of High-Performance Computing (HPC) combined with big data analytics. Fabio Castaldi joined the team to help with project management. ILVO had two demonstrators in the project, for which I coordinated the input into the proposal: one on pig farming and one on open sea fishing. The pig farming one was a combination of advancing the data analysis on my PhD topic and Chari's work in IoF2020, and the advancement of other topics that either colleagues worked on, like hyperspectral analysis of meat for the assessment of quality, or that Vion Food Group, a large Dutch slaughterhouse and meat processor, had an interest in. Open sea fishing was completely new to me, so to be able to follow up this topic taught me a lot of exciting new things!

In between these ongoing projects, Evi and I also took up the challenge of coordinating a project proposal. "aWISH", as we called it, dealt with animal welfare measurements at a very large scale through objective sensor and camera measurements in the slaughterhouse for chickens and pigs. Our hard work was unfortunately not rewarded with an accepted project. We did learn a lot of new skills, however, and although it was not accepted, I still believe we did a very good job for a first time, and we continue to look for other funding opportunities.

Today, we are approaching the last year of IoF2020, and CYBELE is one-third of the way through. I now have the privilege of directly supervising two PhD students and three researchers and project managers. I really enjoy following this challenging process from another angle this time. I'm currently also taking over another project, SmartAgriHubs, from a colleague that is on maternity leave. This brings me back to the day-to-day work of project management, a lot of telcos, writing minutes, and solving issues while monitoring progress and producing deliverables. And this time, I just enjoy it because I know I have the skills to do it!

9.7   Looking Back and Vision on the Future

The main theme among my work still remains pig farming. From 8 years ago, when I started my PhD with no knowledge about pigs at all, I now have two pet pigs at home on pasture and am the contact person for technology in pig farms at ILVO. No doubt I still have a lot to learn about pig farming. However, my experience shows that you should never think that you cannot do a certain thing because you have a different degree or a different background. Never be too shy to try something new if you want to and to follow your dreams.

Where will I be in 8 years? I don't know, as long as I feel fulfillment in the job that I do and that there is enough flexibility and variation, I will be fine. I hope to see the topics and projects that I have been working on develop further and in the end contribute to practice, being used in pig farms and making an impact. Also, I hope to make an impact on the life of the pigs by being able to detect if they are having health or welfare issues and do something about it while creating awareness for their individual needs.

Acknowledgments I acknowledge the help of all my colleagues, supervisors, project partners, and other professionals that I came across in my career so far for shaping my professional experiences and giving me all the opportunities I had so far. In addition, I want to thank my family, friends, and especially my husband Steven Lecompte for their support in my private life. Also, thanks to Steven and Editor Takoi Hamrita for improving my manuscript.

References

Heitkämper, K., Schick, M., & Fritzsche, S. (2011). Working-time requirement in pig fattening. *Landtechnik, 66*, 113–115.
Maselyne, J. (2016). *Automated monitoring of feeding and drinking patterns in growing-finishing pigs: Towards a warning system for performance, health and welfare problems in individual pigs*. PhD dissertation in bioscience engineering, KU Leuven, Belgium.

Maselyne, J., Saeys, W., De Ketelaere, B., et al. (2014a). Validation of a High Frequency Radio Frequency Identification (HF RFID) system for registering feeding patterns of growing-finishing pigs. *Computers and Electronics in Agriculture, 102*, 10–18.

Maselyne, J., Van Nuffel, A., De Ketelaere, B., et al. (2014b). Range measurements of a High Frequency Radio Frequency Identification (HF RFID) system for registering feeding patterns of growing-finishing pigs. *Computers and Electronics in Agriculture, 108*, 209–220.

Maselyne, J., Adriaens, I., Huybrechts, T., et al. (2015a). Chapt. 5.5: Assessing the drinking behaviour of individual pigs using RFID registrations. In I Halachmi (Ed.), *Precision livestock farming applications: Making sense of sensors to support farm management* (pp. 209–216). Wageningen: Wageningen Academic Publishers.

Maselyne, J., Adriaens, I., Huybrechts, T., et al. (2015b). Measuring the drinking behaviour of individual pigs housed in group using radio frequency identification (RFID). *Animal, 10*(9), 1157–1166.

Maselyne, J., Saeys, W., & Van Nuffel, A. (2015c). Review: Quantifying animal feeding behaviour with a focus on pigs. *Physiology & Behavior, 138*, 37–51.

Maselyne, J., Saeys, W., Briene, P., et al. (2016). Methods to construct feeding visits from RFID registrations of growing-finishing pigs at the feeder trough. *Computers and Electronics in Agriculture, 128*, 9–19.

Maselyne, J., Van Nuffel, A., Briene, P., et al. (2018). Online warning systems for individual fattening pigs based on their feeding pattern. *Biosystems Engineering, 173*, 143–156.

Statistics Flanders (2019). Accessed online at Dec 2019. https://www.statistiekvlaanderen.be/land-en-tuinbouwbedrijven

van den Heuvel, E. M., Hoofs, A. I. J., Binnendijk, G. P., et al. (2004). *Grote groepen vleesvarkens. Effects of group size on fattening pigs* (29th ed.). Lelystad: Wageningen UR.

Dr. Jarissa Maselyne, PhD MSc (female), has a Master of Science in Electromechanical Engineering with a major in Control Engineering and Automation (University of Ghent, 2011). In 2016, she obtained a PhD degree in Bioscience Engineering at the KU Leuven with the topic "Automated monitoring of feeding and drinking patterns in growing-finishing pigs: towards a warning system for performance, health and welfare in individual pigs." She is working as a researcher and project manager at ILVO (Flanders Research Institute for Agriculture, Fisheries, and Food) in the group of Agricultural Engineering, with the main focus on precision pig farming. She has been involved in several international projects such as PigWise (ICT-Agri Era-net funded), IOF2020, CYBELE, and SmartAgriHubs (all H2020 funded). She is a team leader for several projects and for the precision pig farming group at ILVO. She is also vice president of the precision livestock farming committee at EAAP (European Federation of Animal Science) and is an editor for Animal, a Cambridge international journal of animal bioscience.

Chapter 10
The Impact of Advances and Challenges of Bush Internet Connectivity for Women in Agriculture in Queensland, Australia

Rachel Hay

Contents

Topics: Women in ag, connectivity, access, reality and implication of available precision ag tools, remote use of Internet (health, well-being), how tech can assist with succession, what does a sustainable future look like for AU women in ag.

10.1 Biography

I am not afraid of storms, for I am learning to sail my ship (Aeschylus, 525 BC–456 BC).

I was a late academic bloomer. For as long as I can remember, I wanted to grow up and be a mom, I knew that once I had reared my children, only then could I fulfill

---

R. Hay (✉)
James Cook University, Townsville, Australia
e-mail: rachel.hay@jcu.edu.au

© Springer Nature Switzerland AG 2021
T. K. Hamrita (ed.), *Women in Precision Agriculture*, Women in Engineering and Science, https://doi.org/10.1007/978-3-030-49244-1_10

my own needs. I always knew I would "go back to school," but I was not sure of what that looked like. I was very entrepreneurial. I always had a job, owned my own business, and, with my husband, was happy to build our empire. That was until our best friends died within a year of each other. Following the advice of Robert Kiyosaki (2000) – if you don't know what to do, take a year off and let your brain think without pressure – we took a year off (2008/2009) to drive (with the kids) around Australia in a caravan. It was on return from this trip that I decided to enroll at university; I was 39 years old.

I enrolled in a business degree because I was particularly interested in regional development, and while our local university offered the pathway, it meant I had to take economics. Taking economics was very daunting, I had not done very well in high school, and it had been 20 years since I had looked at education, let alone the complex mathematical problems that came with economic modeling. As a first in family university student, I was determined to win the battle and show my children that education was cool. So through persistence and tears and with a very supportive husband, I forged on. About halfway into my degree, I realized that the economics degree I was taking was heading toward banking, which I was not interested in, so I changed my major to marketing. It was at this point I realized my passion for behavior change and social marketing and how the discipline can be used in regional development, and now 10 years on, I wear a lovely regional development/social marketing hat and channel my energy towards creating change.

After doing well in my undergraduate, I started an honors degree where my love of research and agriculture combined to create change in rural, regional, and remote (RRR) Australia. Not only were RRR families suffering from the effects of a digital divide and a paucity of Internet connectivity (Correa and Pavez 2016); women were struggling to keep their children's education at the same level as the city kids and to break into a new world outside of motherhood, farmhand, and homemaking. Another thing that really stood out was that rural digital technology, used readily in other agricultural industries, was not being used in the beef cattle industry and that this was not solely due to a lack of access or connectivity issues. A situation worth studying, and so my first thesis was born.

My research allowed me to develop a reality among rural people through interpreting their meanings and understandings during conversational research and to share that reality with wider audiences. The respondents' subjective reality allowed me to truly understand how a lack of Internet connectivity affected RRR business, education, and personal well-being, contributing to hardship in RRR areas, and how this affected both women and men farmers. While the study was about women's adoption and use of technology, it is important to remember that for Australian farming families, many of the RRR women and men work as a team, and they are not always cognizant of a gender divide. Therefore, the most important lesson from the research was to acknowledge the farming partnership and what effect the paucity of technology connectivity had on farming relationships through the lenses of women.

10.2   The Beginning of Behavior Change for Rural Digital Technology

For 3 days, I walked around a dusty agricultural field day interviewing Australian women and men (60 face-to-face interviews) about their technology use. It turns out that the men interviewed did not think that they could use technology and that the interviewed women, who were using technology because they had to, were becoming experts. Three years later, I repeated my study, completing face-to-face interviews again to find that there were some changes in how rural digital technology was being used (more about this later) (Fig. 10.1).

My study reignited previous work on women in agriculture in Australia, and change started to happen. For example, one of the digital technology (water sensors) providers involved in my study changed their marketing strategy to focus on women as decision-makers, significantly increasing their sales. Another example includes where men decided that technology may be easy to learn and use and, as such, started using their smartphones to purchase spare parts, research markets, and check the weather. Rural digital technology in the study was identified as computers, laptops, smartphones, satellite phones, tablets, and walk over weighing systems, water

Fig. 10.1 Field day handbook

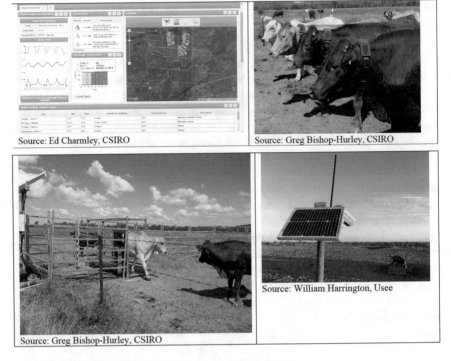

Source: Ed Charmley, CSIRO

Source: Greg Bishop-Hurley, CSIRO

Source: William Harrington, Usee

Source: Greg Bishop-Hurley, CSIRO

Fig. 10.2 Examples of rural digital technology (Usee, CSIRO)

sensor technology, livestock theft technology, and the National Livestock Identification System (NLIS), apps, and paddock to plate management systems (see some examples in Fig. 10.2).

10.3 A Historical View of Women's Technology Adoption

When I first started my PhD journey, I found that there were many articles about technology adoption by women entrepreneurs, women in construction, and the socioeconomic status of women, as well as strategies for empowerment of women through adoption of technology in the rural home (3BL Blogs 2014; Anugwom 2011; Ndubisi 2007; Verma et al. 2013). However, very little research was available that identified factors that influenced women's decision to adopt rural digital technology – there is still a very large gap in the research here – and there is also very little historical evidence about women's motivations for adopting technology of any kind. This was intriguing and spurred me on to continue to dig deeper.

By chance, I found an older article by Cowan (1979) that identified women "as bearers of children" (p. 52). Cowan (1979) highlighted that there are many inventions that are aimed at women, for example, the baby bottle, sterilizers, childbirth

Fig. 10.3 Advertising depicting women using technology designed for women and for men ((left) Bruce Floor Products 1948, (right) Mowa-Matic Lawn Mower 1953)

interventions, and the baby carriage among other inventions such as the washing machine and vacuum cleaner, highlighting women's role as being in the household (Fig. 10.3).

This research trend – of women only being recognized for domestic duties – continued well into the 2000s (Alston and Wilkinson 1998; Bryant and Pini 2006; Little 2009; Whatmore 1991). Early advertising often depicted women doing both women's and men's duties, but rarely (if ever) the opposite, where men were doing women's duties.

As I worked through the historical literature, it discussed male and female roles in purchase decisions. It highlighted males as the dominant purchasers of "brown goods," i.e., goods that are for the home but not for housework, for example, the stereogram, and females as the decision-makers for purchases of "white goods," i.e., goods that are used for housework, for example, the washing machine (Bose et al. 1984). However, men rather than women were identified as the drivers of those purchase decisions. For the most part, women made decisions in the household about food, small appliances, and "purchases that reflect female activity in the home" (Bose et al. 1984, p. 61). Men were responsible for the remaining purchase decisions, and typically, women were measured against the practices of men's work and men's lives as the norm (Little 2009).

In this study, while men and women were identified as joint decision-makers, this depended on the size of the purchase decision, i.e., if it was high cost, then the decision was made together in most cases (93.0%). However, decisions about inside and outside technology were still gender driven, for example, 81.9% of women were identified as the decision-makers if the decision was to be made about an inside (e.g., computer, accounting, cattle management, or education program) purchase. When the purchase was for outside equipment (e.g., walk over scales, remote weather stations), 76.9% of respondents agreed that it was the man's decision.

Very little historical writing about women's approach to the adoption of these or any other technology exists, reinforcing how invisible women farmers have been (and possibly still are). The term "invisible farmer" is increasingly being used to define women who work in agriculture where their role is difficult to describe (Brandth 2002). Women's position in farming is often tied to their marital contract, seen as the farmer's wife without independent status (Brandth 2002, p. 184). The farm work that women do is often overlooked, unnoticed, and invisible to others (Brandth 2002; Little 1987; Whatmore 1991), hence the term "invisible farmer." To contrast this, among others in Australia, work is being completed by Museums Victoria as part of the Invisible Farmer Project funded by an Australian Research Council Grant (Museums Victoria 2017), which aims to recognize women's contribution to agriculture.

10.4 The Gender Divide

The gender division in agriculture highlighted a social inequality that is not only based on gender (Little 2009) but also based on gender relations, capital resources (women marry into the family so men own the capital), and decision-making (Bock 2006). The results from my survey indicated that 38% of women were identified as "invisible farmers" (Brandth 2002; Bryant and Pini 2006), labelling themselves as wives, spouses, or daughters of the property. This gender division brings a focus to the disadvantages of rural residency (suffering problems of remoteness and poor services) and a lack of social and economic equality for women (Little 2009; Penley 1991).

There is also a link between gender and technology where men using agricultural technology reinforce patriarchal ideologies that ultimately marginalize women to exclude them from both farming and decisions about farming (Bryant and Pini 2006; Saugeres 2002). This link is challenged by this research. These marginalizations were justified by the proposition that women's changing role on-farm may challenge men's masculinity as men associate working with machinery with leadership (Brandth 1995, 2006). Previously, by downplaying the gender difference, the essential role of women in economic profitability of and therefore viability of the farm could essentially be ignored (Little 2009). However, according to farming women in Australia, men are not challenged by women's adoption of technology.

The results from this study showed that more women reported using typically male-oriented rural technology. For example, 44.5% of women reported using NLIS management systems compared to men (31%), and there was an increase in women using technology such as satellite mapping, remote cameras, in vitro fertilization (IVF), and GPS property management tools (Hay 2018). When asked how valuable women working with technology was to the farming business, 94.7% of men responded that it was an important contribution. They also commented that women keeping up with technology allow better decision-making and that their contribution makes them a valuable part of the team.

While women worked to contribute to agricultural production, they also combine off-farm, nonfarm employment or other revenue streams (Blad 2012; Eikeland and Lie 1999). This was not seen as a threat to masculinity but encouraged as a contribution to diversifying income. Over time, farming women have become more involved in decision-making, which has led to women's role in farming being recognized as valuable (Alston and Wilkinson 1998; Claridge 1998; Farmar-Bowers 2010; Gasson and Winter 1992; Pannell and Vanclay 2011; Rickson and Daniels 1999; Umrani and Ghadially 2003).

10.5 Australian Women in Agriculture

Women involved in Australian beef production suggest that there is very little division of gender within the industry. Rural women see women and men as being on a level playing field: "It's about being together, men and women on farms don't see themselves as having a gender based role, but as the job getting done by whomever is available to do it" (Interview number 8) (Fig. 10.4).

These views are being recognized in recent research; for example, Beach (2013) states: "while the discourses of the family farm and masculinization do occur...neither one is the primary discourse expressed by farmers" (p. 225), supporting Australian rural women's views. However, the recognition is not yet far-reaching.

Australian women surveyed in the study worked on or owned properties that are between 10,000 and 20,000 acres in size (76%), they primarily worked cattle (96.5%), and the majority had children either at home or at boarding school (which means they came home part-time). Australian women in agriculture have not always worked in the industry. The majority of women in the survey were previously teachers and worked in education, administration, or nursing, closely followed by banking, finance, or commerce. They used technology both within the home and in the paddock (laptop/tablet, National Livestock Identification System (NLIS), remote cameras, remote weather stations, walk over scales, remote water cameras and sensors, and in vitro fertilization (IVF) technology). Women participate in activities such as online banking (85.9%), accounting (85.0%), business (72.8%), personal communication (emails 72.1%), social media (62.7%), and communicating with friends and neighbors (70.7%). Social media was used (2016) in many cases to sell cattle and other farm products (machinery, fencing, etc.) and to run their online businesses from their remote cattle stations.

While women are embracing technology, it was not always their first choice; for many, the use of technology fell to them because they worked primarily from the

Fig. 10.4 Australian women in agriculture

house. However, as technology becomes more mobile, others on the farm are becoming users as well. In the first round (2013) of data collection, none of the men in the farming family were using technology. By the second round of data collection, 3 years later (2016), "others" on the farm were using some sort of technology; this in turn was reducing some of the technology user burden from the farming women. The women identified the new users as their husbands (the highest technology users, 22% more than the first year), followed by both male and female workers and children. When making decisions about technology purchased for the farm, the results show that men and women were making them together in most cases (73.9%) in the first round of study. By the second round, women were making more decisions about technology on their own (43.7%), indicating a shift toward women making decisions on farm.

10.6 Technology and Well-Being

Australian women in the study highlighted that technology was also helping to reduce isolation (75.4%), and one woman respondent quoted, "Having access to information decreases social isolation, access to family and friends via Skype is good for my mental health" (Hay 2018, p. 245). Australia has a high rate of suicide in RRR areas (Roy et al. 2017). The mental health and well-being of people in rural areas of Queensland have suffered from the effects of prolonged drought and other external factors (such as the interruption to live meat export trade (2011) and more recently (2019) extensive floods and large-scale bushfires in 2020). For example, live meat export from Australia worth $1.4 billion was suspended from June 2011 for 6 months due to cases of animal cruelty being exposed in Indonesian abattoirs, halting trade and devastating farmers and regional economies (McDonald et al. 2011; Wagstaff 2016). The consequences from flooding and bushfires have not been fully quantified at the time of writing.

Technology has extended resources and access to services; as well, it has increased training opportunities for health workers, which has a positive effect on mental health and well-being (Allan 2010). Women who can access health and well-being programs online can use the information to help the men in their family (Powell et al. 2012), whether they be husbands, fathers, sons, or workers, to access well-being services. Women's access to male family members and workers has been identified as an entry point for male-related mental health and may well be the key to increased well-being, especially in men: "Women are the key to accessing men because women are crucial to keeping families together" (Congues 2014). Men's responses agree highlighting that women's role in technology helps to speed things up, leaving more time to do things as a family, highlighting that "people really have no idea how important this [having more time for family] is to people in rural Australia" (Hay 2018, p. 161). In addition, access to technology means that workers, including adult children, are happier on farming properties (79.4%). However, this happiness does not totally translate to women's mental health.

While mental health was mostly self-rated as positive, 43.2% of women self-rated their mental health as moderate, 6% self-rated it as poor or extremely poor, and 29.6% self-rated their mental health as excellent. The moderate scores may be due to some women not wanting to be responsible for using technology (47.4%) as working with technology ignores competing priorities that ask women to decide between other duties that they perform (such as off-farm work) and technology-based duties (most often delegated to night work). Women are sometimes frustrated being responsible for using technology and then getting the blame when something goes wrong, adding to their frustration. While women reported that working off-farm had little effect on their emotional health, they indicated that working off-farm may not be satisfying as they would rather be working on farm. Schirmer et al. (2018), Chang and Mishra (2008), McCoy and Filson (1996), and Van den Broeck and Maertens (2017) support this comment as the researchers found that women who work off-farm may be less satisfied than those who only work on-farm or do not work at all.

The majority of female and male participants agreed that having access to the Internet increases their quality of life/well-being and that by having access to the Internet, they feel more equal to people in more Internet-accessible areas. Participants agreed that completing computer-based work at night allowed the women to work alongside their partners during the day, which made their partner happier (less suicidal) and, as such, made their life happier. They also agreed that being able to work alongside others during the day contributed to their sense of well-being. Female respondents agreed that partners', children's, and workers' mental health was positively affected by the woman working outside during the day and on the computer at other times.

10.7   Bush Internet Challenges and Advances

Australia is a big country; it covers approximately 7.692 million square kilometers (Dept. of Environment and Energy 2005). Around 81% of this is broadly defined as encompassing rural, regional, and remote areas of Australia. Rangelands are characterized by "eucalypt savanna and native grasslands, small areas of cleared land and scattered settlements, and rivers and wetlands that sustain ecosystems" (Dept. of Environment and Energy 2005, p. 1). The rangelands are the home of Australian beef cattle production, where the industry produces around 27 million head of cattle to the value of $17.87 billion (2014/15) (Meat and Livestock Australia 2016). Agricultural products from Australia are highly regarded, as such the beef industry looks to Information Communication Technology (ICT) to help boost production to meet projected global food demand goals (Linehan et al. 2012).

ICTs have the potential to transform how people live in rural, regional, and remote areas. New Internet and mobile phone technology is allowing producers to keep in contact, not only with friends and relatives but also with markets, suppliers, telehealth services, weather, flood and fire services, and banking as well as remote education. However, access to networks in Australian rangelands is challenging (Hay 2016; Curtin 2001). ICT connectivity across the rangelands is limited. For example, Fig. 10.5 demonstrates a lack of overlap between the Australian rangelands and Internet responsiveness to a national survey of Internet usage (Hay 2016). In many sparsely populated pastoral regions, download speeds can be as low as 0.7 Mbps (Hay 2016). Expensive and unattainable access to either mobile or Internet connectivity is adding to the digital divide currently experienced by those on the rangelands (Curtin 2001). My research supports a small group of cattle-producing women who are using social marketing to advocate for change.

As my journey progressed, my work in technology adoption by women in agriculture led me to a volunteer group of Australian cattle-grazing women who had started an advocacy group aimed at #fixingbushinternet in RRR Australia. The team employs social marketing practices and advocacy (Novelli 2011) and uses community connectedness through their social networks (Lefebvre 2013) to focus on fixing bush Internet and putting an end to the data drought experienced by RRR communities in Australia. I became their data analyst and presented their survey data to a level that was recognized by the government (Hay 2017). We as a team wrote submissions to parliamentary inquiries to advocate for better and fairer Internet in the bush. We lobbied government, relevant industry bodies, and telecommunications providers to highlight shortfalls in service provision. These shortfalls would have

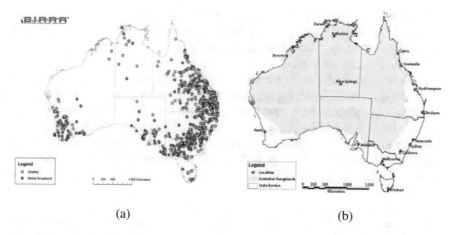

(a)   (b)

Fig. 10.5 (**a**) Responses mapped from the Better Internet for Rural Regional and Remote Australia (BIRRR) Regional Internet Access Survey, showing access to Internet in Australia (Kristy Sparrow, BIRRR Regional Internet Access Survey Results 2016). (**b**) Map of Australian rangelands. (Source: Commonwealth of Australia, Australian Government Department of the Environment and Heritage 2005)

previously gone unnoticed. The group, through widespread surveys and advocacy, has successfully achieved unmetered access to specific distance education sites on Telstra mobile broadband, dedicated education ports on NBN™[1] Sky Muster, a dedicated RRR contact team with NBN™, and increased data plans on NBN™, each of these decreasing the digital divide in RRR areas of Australia.

10.8   Women's Motives, Actions, and Intentions for Technology Use and Management

Women are using technology more than men, and when men are using technology, it has usually been purchased, set up, and maintained by women. Specifically, technology used by women relates to both management practices and social connection. Both management practices and social connection are leading to less isolation in terms of having better access to business management and communication tools. Women are motivated to use technology to research and improve production, manage accounting practices, and improve communication to create opportunity for and within their family business.

Interestingly, while men are using technology more in 2016 than in 2013, they are using the technology-based tools (nearly always installed by women) more

---

[1] NBN Co Limited is an Australian government-owned corporation tasked to design, build, and operate Australia's National Broadband Network as a monopoly wholesale broadband provider.

practically, for example, by checking market pricing, weather, remote sensing technology, and remote cameras; saving time by streamlining farming systems; and increasing productivity rather than seeking decision-making information.

Women and men are making decisions about technology together, whereas in the past, this was reported as being the men's role. Although some women are frustrated by having extra duties involving technology, they are motivated to continue to manage technology. Significantly, women are gaining valuable skills over time, and they feel a sense of achievement, empowerment, and self-worth by managing technology. Importantly, having technology on the property has improved management practices, decision-making, record keeping, and planning, and it has given access to information that women would otherwise not be able to gather while situated at their property.

Technology can be intrusive. However, it can also save time and money and allow for broader communication with peers, industry, and customers, increasing competitiveness in their industry. The study supports earlier research that reduced isolation and the depletion of the tyranny of distance (Blainey 1983) are encouraging women to adopt technology in rural, regional, and remote areas of Australia. Therefore, a strong focus on increasing Internet connectivity by government and other stakeholders is required.

The Technology Acceptance Model (TAM) (Davis 1989) is supported by this study as a practical application that suits the personality styles of women producers, namely, that it uses both perceived ease of use and perceived usefulness to determine a person's attitude toward adoption. Adoption is determined by each woman's individual environment, which will be affected by other factors, for example, technological, socioeconomic, agro-economic, institutional, informational, and behavioral factors, and by producers' perceptions (Tey and Brindal 2012). Producer decisions about adoption will be driven by the problem at hand (Öhlmér et al. 1998); by cognitive and normative influences (Bearden and Etzel 1982; Miller et al. 2011) as well as by family and individual motivation, suitability, and opportunities available (Farmar-Bowers 2010); and by a producer's level of adaptive capacity (Berry et al. 2011). However, while a woman's adaptive capacity can give her the ability to change and take advantages of opportunities or cope with stress, ongoing challenges with Internet connection for rural, regional, and remote Australia still present as the biggest barrier to technology adoption.

Technology that is more portable, such as laptops, smartphones, and tablets, are being used on farm, demonstrating that technology is being used outside of the homestead. Women are using both practical and communication technology but are moving away from things like searching on the Internet toward using social media to run off-farm businesses or to sell their cattle. While men are using technology more, women are still responsible for purchasing, programming, and teaching male producers how to use the selected technology. Having access to the Internet is increasing quality of life for both women and men as well as children and workers. While some women are still somewhat reluctant to take on technology on-farm, others feel empowered and valued in their work. As technology is diffusing into rural settings, it is modifying gender divisions and supporting women as they move from traditional separate roles in decision-making to productive partnerships in farming families, encouraging stakeholders to see women as both decision-makers and community leaders.

10.9 Contribution to the Topic

This chapter contributes to existing knowledge about diffusion of rural digital technology into beef-producing families. It informs government, policy makers, and other stakeholders including the media and communications and technology service providers about factors that influence technology adoption and women's key role in adoption decisions and thus how to enable rural women to support their farming business and their lifestyle as well as increase productivity in the beef industry. It highlights women's role in decision-making in beef production practices, identifying how digital technology affects the beef production business, personal career path, and family aspirations from a women's perspective.

The work recognizes the importance of the women's role in decision-making in beef production practices, increasing self-worth and importance. Recognition of women as producers may help to shift their roles from representatives of the beef industry to participants in decision-making about the beef industry, allowing rural women to build networks and contribute to the beef-producing community. However, the implications for the research reach beyond the beef-producing community to benefit the wider community by providing food security, jobs, tourism opportunities, and a future for beef producer's children.

10.10 Conclusion

I am immensely proud of my work over the past 9 years in technology adoption by women in agriculture. It has led me on a pathway that was previously unknown to me. I have been able to apply the knowledge I have gained to other areas including best management practice in water quality on the Great Barrier Reef, business coaching for rural producers, widening participation in education, advocacy groups, behavior change, and readability and communications. I have produced 34 publications (academic and industry) over the past 5 years, and I have recently been elected as a Board Member for the Queensland Rural Regional and Remote Women's network, which aims to connect, develop, and inspire RRR women. My research has allowed me to travel to present at many reputable national and international conferences, even those outside of my comfort zone. As a social scientist, I recently presented findings from the behavior change project that aims to understand farmers' best management nutrient practices to distinguished, world-renowned expert scientists (you know the real ones), which was daunting, but I did it! I bring with me my own thoughts and biases, and I respect those of others. As I travel through these next few years as an early career researcher, I hope to continue to develop my niche and my expertise and to enjoy the behavior change of a nation as it traverses the digital divide and all that technology adoption can bring with it.

References

Allan, J. (2010). Determinants of mental health and well-being in rural communities: Do we understand enough to influence planning and policy? *Australian Journal of Rural Health, 18*(1), 3–4. https://doi.org/10.1111/j.1440-1584.2009.01121.x.

Alston, M., & Wilkinson, J. (1998). Australian farm women – Shut out or fenced in? The lack of women in agricultural leadership. *Sociologia Ruralis, 38*(3), 391–408. https://doi.org/10.1111/1467-9523.00085.

Anugwom, E. E. (2011). Adoption of technology and the socio-economic status of rural women in South-Eastern Nigeria. *Africa Insight, 41*(3), 16–29.

Beach, S. S. (2013). "Tractorettes" or partners? Farmers' views on women in Kansas farming households. *Rural Sociology, 78*(2), 210–228. https://doi.org/10.1111/ruso.12008.

Bearden, W. O., & Etzel, M. J. (1982). Reference group influence on product and brand purchase decisions. *Journal of Consumer Research, 9*(2), 183–194.

Berry, H. L., Hogan, A., Owen, J., Rickwood, D., & Fragar, L. (2011). Climate change and farmers' mental health: Risks and responses. *Asia-Pacific Journal of Public Health, 23*(2 suppl): 119S–132S.

Blainey, G. (1983). *The tyranny of distance: how distance shaped Australia's history* (Rev. ed ed.). Melbourne VIC: Sun Books.

Blad, M. (2012). *Pluriactivity of farming families – old phenomenon in new times*. Institute of Rural and Agricultural Development. Poland. Retrieved from http://ageconsearch.umn.edu/bitstream/139799/2/vol.%207_12.pdf

Bock, B. S (2006). *Rural gender relations: Issues and case studies*. Retrieved from http://site.ebrary.com.elibrary.jcu.edu.au/lib/jcu/docDetail.action?docID=10255055

Bose, C. E., Bereano, P. L., & Malloy, M. (1984). Household technology and the social construction of housework. *Technology and Culture, 25*(1), 53–82. https://doi.org/10.2307/3104669.

Brandth, B. (1995). Rural masculinity in transition: Gender images in tractor advertisements. *Journal of Rural Studies, 11*(2), 123–133.

Brandth, B. (2002). Gender identity in European family farming: A literature review. *Sociologia Ruralis, 42*(3), 181–200. https://doi.org/10.1111/1467-9523.00210.

Brandth, B. (2006). Agricultural body-building: Incorporations of gender body and work. *Journal of Rural Studies, 22*(1), 17–27. https://doi.org/10.1016/j.jrurstud.2005.05.009.

Bryant, L., & Pini, B. (2006). Towards an understanding of gender and capital in constituting biotechnologies in agriculture. *Sociologia Ruralis, 46*(4), 261–279. https://doi.org/10.1111/j.1467-9523.2006.00417.x.

Chang, H.-H., & Mishra, A. (2008). Impact of off-farm labor supply on food expenditures of the farm household. *Food Policy, 33*(6), 657–664. https://doi.org/10.1016/j.foodpol.2008.02.002.

Claridge, C. (1998). Rural women, decision making and leadership within environmental and landcare groups [online]. *Rural Society, 8*(3), 183–195.

Congues, J. (2014). Promoting collective well-being as a means of defying the odds: Drought in the Goulburn Valley, Australia. *Rural Society, 23*(3), 229–242.

Correa, T., & Pavez, I. (2016). Digital inclusion in rural areas: A qualitative exploration of challenges faced by people from isolated communities. *Journal of Computer-Mediated Communication, 21*(3), 247–263. https://doi.org/10.1111/jcc4.12154.

Cowan, R. S. (1979). From Virginia dare to Virginia slims: Women and technology in American life. *Technology and Culture, 20*(1), 51–63. https://doi.org/10.2307/3103111.

Curtin, J. D. (2001) *A Digital Divide in Rural and Regional Australia?* Current Issues Brief.

Davis, F. D. (1989). Perceived usefulness, perceived ease of use, and user acceptance of information technology. *MIS Quarterly, 13*(3), 319–340.

Dept. of Environment and Energy. (2005). *Introduction to Australia's rangelands. Outback Australia - the rangelands*. Retrieved 2 January, 2017, from https://www.environment.gov.au/land/rangelands.

Eikeland, S., & Lie, I. (1999). Pluriactivity in rural Norway. *Journal of Rural Studies, 15*(4), 405–415. https://doi.org/10.1016/S0743-0167(99)00010-8.

Farmar-Bowers, Q. (2010). Understanding the strategic decisions women make in farming families. *Journal of Rural Studies, 26*(2), 141–151. https://doi.org/10.1016/j.jrurstud.2009.09.008.

Gasson, R., & Winter, M. (1992). Gender relations and farm household pluriactivity. *Journal of Rural Studies, 8*(4), 387–397. https://doi.org/10.1016/0743-0167(92)90052-8.

Hay, R. (2016). *Better Internet for Rural, Regional and Remote Australia: Regional Internet Access Survey Results, 2016*. Retrieved from https://birrraus.files.wordpress.com/2017/04/birrr-report-2016-survey-resultsfinal.pdf.

Hay, R. (2017). *Better Internet for Rural, Regional and Remote Australia: Skymuster Survey Results 2017*. Retrieved from https://birrraus.files.wordpress.com/2017/04/birrr-report-2016-survey-results-final.pdf.

Hay, R. (2018). *The Engagement of Women and Technology in Agriculture.* Doctor of Philosophy, Management, and Commerce PhD, James Cook University. https://researchonline.jcu.edu.au/53136/. https://doi.org/10.4225/28/5ad015a60c689

Kiyosaki, R. (2000). *Rich Dad, Poor Dad: What The Rich Teach Their Kids About Money*: Scribl.

Lefebvre, R. C. (2013). *Social marketing and social change: Strategies and tools for improving health, well-being, and the environment*. John Wiley & Sons.

Linehan, V., Thorpe, S., Andrews, N., eon, K., & Beaini, F. (2012). *Food Demand to 2050: Opportunities for Australian agriculture*. ABARES Outlook Conference. F. a. F. Department of Agriculture, ABARES. Canberra, ACT, Australian Government.

Little, J. (1987). Gender relations in rural areas: The importance of women's domestic role. *Journal of Rural Studies, 3*(4), 335–342. https://doi.org/10.1016/0743-0167(87)90052-0.

Little, J. (2009). Gender and Rurality. In Editors-in-Chief: K. Rob & T. Nigel (Eds.), *International Encyclopedia of human geography* (pp. 315–319). Oxford: Elsevier.

McCoy, M., & Filson, G. (1996). Working off the farm: Impacts on quality of life. *Social Indicators Research, 37*(2), 149–163.

McDonald, S., Henderson, A., & Middleton, A. (2011). Industry hit hard by live export ban. *ABC News*. Retrieved from http://www.abc.net.au/news/2011-06-08/industry-hit-hard-by-live-export-ban/2750768

Meat and Livestock Australia (2016). Fast facts: Australia's beef industry. Online at https://www.mla.com.au/Prices-markets/Trends-analysis/Fast-Facts, Meat and Livestock Australia, 2016.

Miller, D., Le Breton Miller, I., & Lester, R. H. (2011). Family and lone founder ownership and strategic behaviour: Social context, identity, and institutional logics. *Journal of Management Studies, 48*(1), 1–25.

Museums Victoria. (2017). *Invisible Farmer Project*. Victoria, Australian: Australian Research Council (ARC).

Ndubisi, N. O. (2007). Evaluating the direct and indirect impact of traits and perceptions on technology adoption by women entrepreneurs in malaysia. *Academy of Entrepreneurship Journal, 13*(2), 1–20.

Novelli, W. D. (2011). *The SAGE Handbook of Social Marketing*. B. Workman. London, SAGE Publications Ltd.

Öhlmér, B., Olson, K., & Brehmer, B. (1998). Understanding farmers' decision making processes and improving managerial assistance. *Agricultural Economics, 18*(3), 273–290.

Pannell, D., & Vanclay, F. (Eds.). (2011). *Changing land management: Adoption of new practices by rural landholders*. Collingwood: CSIRO Publishing.

Penley, C. (1991). Brownian motion: Women, tactics and technology. In C. Penley & A. Roiss (Eds.), *Technoculture* (pp. 139–161). Minneapolis: University of Minnesota Press.

Powell, J., Hamborg, T., Stallard, N., Burls, A., McSorley, J., Bennett, K., et al. (2012). Effectiveness of an internet-delivered cognitive-behavioural aid to improve mental wellbeing: A randomised controlled trial. *The Lancet, 380*(Supplement 3), S3. https://doi.org/10.1016/S0140-6736(13)60359-1.

Rickson, S. T., & Daniels, P. L. (1999). Rural women and decision making: Women's role in resource management during rural restructuring 1. *Rural Sociology, 64*(2), 234–250. https://doi.org/10.1111/j.1549-0831.1999.tb00016.x.

Roy, P., Tremblay, G., Robertson, S., & Houle, J. (2017). "Do it All by myself": A Salutogenic approach of masculine health practice among farming men coping with stress. *American Journal of Men's Health, 11*(5), 1536–1546. https://doi.org/10.1177/1557988315619677.

Saugeres, L. (2002). "She's not really a woman, she's half a man": Gendered discourses of embodiment in a French farming community. *Women's Studies International Forum, 25*(6), 641–650.

Schirmer, N. A., Schwaiger, M., Taylor, C. R. & Costello, J. P. (2018). Consumer response to disclosures in digitally retouched advertisements. *Journal of Public Policy & Marketing, 37*(1), 131–141.

Tey, Y., & Brindal, M. (2012). Factors influencing the adoption of precision agricultural technologies: a review for policy implications. *Precision Agriculture, 13*(6), 713–730.

Umrani, F., & Ghadially, R. (2003). Empowering women through ICT education: Facilitating computer adoption. *Gender, Technology and Development, 7*(3), 359–377. https://doi.org/10.1177/097185240300700303.

Van den Broeck, G., & Maertens, M. (2017). Does off-farm wage employment make women in rural senegal happy? *Feminist Economics, 23*(4), 250–275. https://doi.org/10.1080/13545701.2017.1338834.

Verma, V., Verma, S., & Rani, E. (2013). Strategies for empowerment of women through adoption of science and technology in rural homes. *Annals of Agri Bio Research, 18*(2), 283–289.

Wagstaff, J. (2016). Effects of Australian live export cattle ban to Indonesia still felt by beef producers. *The Weekly Times*. Retrieved from http://www.weeklytimesnow.com.au/agribusiness/cattle/news-story/86016c75dc8d4c3fdd0b071459b30fb1.

Whatmore, S. (1991). Life cycle or patriarchy? Gender divisions in family farming. *Journal of Rural Studies, 7*(1–2), 71–76. https://doi.org/10.1016/0743-0167(91)90043-R.

3BL Blogs (2014). Women drive technology adoption in construction. Women drive technology adoption in construction http://search.proquest.com/docview/1643217005?accountid=16285.

Dr. Rachel Hay is a social scientist and early career researcher teaching marketing for the College of Business Law and Governance at James Cook University, Townsville, Australia. Rachel's research centers on transdisciplinary approaches to sustained behavior change in social marketing and environmental protection interventions. Her projects aimed at changing behavior in technology adoption, reducing sediment and nutrient runoff in sugarcane farming and grazing, the Digital Homestead Project, Instant Feedback Assessment Techniques, Business Coaching, FIFO, and the Creative Industries. Rachel's passion is to support rural, regional, and remote regions to connect and stay connected through reliable, sustained, and affordable Internet connectivity. Should this occur, then sustained regional development will follow.

Index

© Springer Nature Switzerland AG 2021
T. K. Hamrita (ed.), *Women in Precision Agriculture*, Women in Engineering
and Science, https://doi.org/10.1007/978-3-030-49244-1

Printed in the United States
by Baker & Taylor Publisher Services